How To Solve It

A New Aspect of Mathematical Method

G. POLYA
Stanford University

SECOND EDITION

Princeton University Press
Princeton, New Jersey

L.C. Card: 79-160544
ISBN 0-691-02356-5 (paperback edn.)
ISBN 0-691-08097-6 (hardcover edn.)

First Princeton Paperback Printing, 1971
Second Printing, 1973

Printed in the United States of America
by Princeton University Press, Princeton, New Jersey

From the Preface to the First Printing

A great discovery solves a great problem but there is a grain of discovery in the solution of any problem. Your problem may be modest; but if it challenges your curiosity and brings into play your inventive faculties, and if you solve it by your own means, you may experience the tension and enjoy the triumph of discovery. Such experiences at a susceptible age may create a taste for mental work and leave their imprint on mind and character for a lifetime.

Thus, a teacher of mathematics has a great opportunity. If he fills his allotted time with drilling his students in routine operations he kills their interest, hampers their intellectual development, and misuses his opportunity. But if he challenges the curiosity of his students by setting them problems proportionate to their knowledge, and helps them to solve their problems with stimulating questions, he may give them a taste for, and some means of, independent thinking.

Also a student whose college curriculum includes some mathematics has a singular opportunity. This opportunity is lost, of course, if he regards mathematics as a subject in which he has to earn so and so much credit and which he should forget after the final examination as quickly as possible. The opportunity may be lost even if the student has some natural talent for mathematics because he, as everybody else, must discover his talents and tastes; he cannot know that he likes raspberry pie if he has never tasted raspberry pie. He may manage to find out, however, that a mathematics problem may be as much fun as a crossword puzzle, or that vigorous mental work may be an exercise as desirable as a fast game of tennis. Having tasted the pleasure in mathematics he will not forget it easily and then there is a good chance that mathematics will become something for him: a hobby, or a tool of his profession, or his profession, or a great ambition.

The author remembers the time when he was a student himself, a somewhat ambitious student, eager to understand a little mathematics and physics. He listened to lectures, read books, tried to take in the solutions and facts presented, but there was a question that disturbed him again and again: "Yes, the solution seems to work, it appears to be correct; but how is it possible to invent such a solution? Yes, this experiment seems to work, this appears to be a fact; but how can people discover such facts? And how could I invent or discover such things by myself?" Today the author is teaching mathematics in a university; he thinks or hopes that some of his more eager students ask similar questions and he tries to satisfy their curiosity. Trying to understand not only the solution of this or that problem but also the motives and procedures of the solution, and trying to explain these motives and procedures to others, he was finally led to write the present book. He hopes that it will be useful to teachers who wish to develop their students' ability to solve problems, and to students who are keen on developing their own abilities.

Although the present book pays special attention to the requirements of students and teachers of mathematics, it should interest anybody concerned with the ways and means of invention and discovery. Such interest may be more widespread than one would assume without reflection. The space devoted by popular newspapers and magazines to crossword puzzles and other riddles seems to show that people spend some time in solving unpractical problems. Behind the desire to solve this or that problem that confers no material advantage, there may be a deeper curiosity, a desire to understand the ways and means, the motives and procedures, of solution.

The following pages are written somewhat concisely, but as simply as possible, and are based on a long and serious study of methods of solution. This sort of study, called *heuristic* by some writers, is not in fashion nowadays but has a long past and, perhaps, some future.

Studying the methods of solving problems, we perceive another face of mathematics. Yes, mathematics has two faces; it is the rigorous science of Euclid but it is also something else. Mathematics presented in the Euclidean way appears as a systematic, deductive science; but mathematics in the making appears as an experimental, inductive science. Both aspects are as old as the science of mathematics itself. But the second aspect is new in one respect; mathematics "in statu nascendi," in the process of being invented, has never before been presented in quite this manner to the student, or to the teacher himself, or to the general public.

The subject of heuristic has manifold connections; mathematicians, logicians, psychologists, educationalists, even philosophers may claim various parts of it as belonging to their special domains. The author, well aware of the possibility of criticism from opposite quarters and keenly conscious of his limitations, has one claim to make: he has some experience in solving problems and in teaching mathematics on various levels.

The subject is more fully dealt with in a more extensive book by the author which is on the way to completion.

Stanford University, August 1, 1944

From the Preface to the Seventh Printing

I am glad to say that I have now succeeded in fulfilling, at least in part, a promise given in the preface to the first printing: The two volumes *Induction and Analogy in Mathematics* and *Patterns of Plausible Inference* which constitute my recent work *Mathematics and Plausible Reasoning* continue the line of thinking begun in *How to Solve It.*

Zurich, August 30, 1954

Preface to the Second Edition

The present second edition adds, besides a few minor improvements, a new fourth part, "Problems, Hints, Solutions."

As this edition was being prepared for print, a study appeared (Educational Testing Service, Princeton, N.J.; *cf. Time,* June 18, 1956) which seems to have formulated a few pertinent observations—they are not new to the people in the know, but it was high time to formulate them for the general public—: " . . . mathematics has the dubious honor of being the least popular subject in the curriculum . . . Future teachers pass through the elementary schools learning to detest mathematics . . . They return to the elementary school to teach a new generation to detest it."

I hope that the present edition, designed for wider diffusion, will convince some of its readers that mathematics, besides being a necessary avenue to engineering jobs and scientific knowledge, may be fun and may also open up a vista of mental activity on the highest level.

Zurich, June 30, 1956

Contents

More examples

PART II. HOW TO SOLVE IT

PART III. SHORT DICTIONARY OF HEURISTIC

† Contains only cross-references.

† Contains only cross-references.

PART IV. PROBLEMS, HINTS, SOLUTIONS

HOW TO SOLVE IT

UNDERSTANDING THE PROBLEM

First.

You have to *understand* the problem.

What is the unknown? What are the data? What is the condition?

Is it possible to satisfy the condition? Is the condition sufficient to determine the unknown? Or is it insufficient? Or redundant? Or contradictory?

Draw a figure. Introduce suitable notation.

Separate the various parts of the condition. Can you write them down?

DEVISING A PLAN

Second.

Find the connection between the data and the unknown. You may be obliged to consider auxiliary problems if an immediate connection cannot be found. You should obtain eventually a *plan* of the solution.

Have you seen it before? Or have you seen the same problem in a slightly different form?

Do you know a related problem? Do you know a theorem that could be useful?

Look at the unknown! And try to think of a familiar problem having the same or a similar unknown.

Here is a problem related to yours and solved before. Could you use it? Could you use its result? Could you use its method? Should you introduce some auxiliary element in order to make its use possible?

Could you restate the problem? Could you restate it still differently? Go back to definitions.

If you cannot solve the proposed problem try to solve first some related problem. Could you imagine a more accessible related problem? A more general problem? A more special problem? An analogous problem? Could you solve a part of the problem? Keep only a part of the condition, drop the other part; how far is the unknown then determined, how can it vary? Could you derive something useful from the data? Could you think of other data appropriate to determine the unknown? Could you change the unknown or the data, or both if necessary, so that the new unknown and the new data are nearer to each other?

Did you use all the data? Did you use the whole condition? Have you taken into account all essential notions involved in the problem?

CARRYING OUT THE PLAN

Third.

Carry out your plan.

Carrying out your plan of the solution, *check each step*. Can you see clearly that the step is correct? Can you prove that it is correct?

LOOKING BACK

Fourth.

Examine the solution obtained.

Can you *check the result?* Can you check the argument?

Can you derive the result differently? Can you see it at a glance?

Can you use the result, or the method, for some other problem?

Introduction

The following considerations are grouped around the preceding list of questions and suggestions entitled "How to Solve It." Any question or suggestion quoted from it will be printed in *italics,* and the whole list will be referred to simply as "the list" or as "our list."

The following pages will discuss the purpose of the list, illustrate its practical use by examples, and explain the underlying notions and mental operations. By way of preliminary explanation, this much may be said: If, using them properly, you address these questions and suggestions to yourself, they may help you to solve your problem. If, using them properly, you address the same questions and suggestions to one of your students, you may help him to solve his problem.

The book is divided into four parts.

The title of the first part is "In the Classroom." It contains twenty sections. Each section will be quoted by its number in heavy type as, for instance, "section **7.**" Sections **1** to **5** discuss the "Purpose" of our list in general terms. Sections **6** to **17** explain what are the "Main Divisions, Main Questions" of the list, and discuss a first practical example. Sections **18, 19, 20** add "More Examples."

The title of the very short second part is "How to Solve It." It is written in dialogue; a somewhat idealized teacher answers short questions of a somewhat idealized student.

The third and most extensive part is a "Short Dictionary of Heuristic"; we shall refer to it as the "Dictionary."

It contains sixty-seven articles arranged alphabetically. For example, the meaning of the term HEURISTIC (set in small capitals) is explained in an article with this title on page 112. When the title of such an article is referred to within the text it will be set in small capitals. Certain paragraphs of a few articles are more technical; they are enclosed in square brackets. Some articles are fairly closely connected with the first part to which they add further illustrations and more specific comments. Other articles go somewhat beyond the aim of the first part of which they explain the background. There is a key-article on MODERN HEURISTIC. It explains the connection of the main articles and the plan underlying the Dictionary; it contains also directions how to find information about particular items of the list. It must be emphasized that there is a common plan and a certain unity, because the articles of the Dictionary show the greatest outward variety. There are a few longer articles devoted to the systematic though condensed discussion of some general theme; others contain more specific comments, still others cross-references, or historical data, or quotations, or aphorisms, or even jokes.

The Dictionary should not be read too quickly; its text is often condensed, and now and then somewhat subtle. The reader may refer to the Dictionary for information about particular points. If these points come from his experience with his own problems or his own students, the reading has a much better chance to be profitable.

The title of the fourth part is "Problems, Hints, Solutions." It proposes a few problems to the more ambitious reader. Each problem is followed (in proper distance) by a "hint" that may reveal a way to the result which is explained in the "solution."

We have mentioned repeatedly the "student" and the "teacher" and we shall refer to them again and again. It may be good to observe that the "student" may be a high school student, or a college student, or anyone else who is studying mathematics. Also the "teacher" may be a high school teacher, or a college instructor, or anyone interested in the technique of teaching mathematics. The author looks at the situation sometimes from the point of view of the student and sometimes from that of the teacher (the latter case is preponderant in the first part). Yet most of the time (especially in the third part) the point of view is that of a person who is neither teacher nor student but anxious to solve the problem before him.

How To Solve It

PART I. IN THE CLASSROOM

PURPOSE

1. Helping the student. One of the most important tasks of the teacher is to help his students. This task is not quite easy; it demands time, practice, devotion, and sound principles.

The student should acquire as much experience of independent work as possible. But if he is left alone with his problem without any help or with insufficient help, he may make no progress at all. If the teacher helps too much, nothing is left to the student. The teacher should help, but not too much and not too little, so that the student shall have a *reasonable share of the work.*

If the student is not able to do much, the teacher should leave him at least some illusion of independent work. In order to do so, the teacher should help the student discreetly, *unobtrusively.*

The best is, however, to help the student naturally. The teacher should put himself in the student's place, he should see the student's case, he should try to understand what is going on in the student's mind, and ask a question or indicate a step that *could have occurred to the student himself.*

2. Questions, recommendations, mental operations. Trying to help the student effectively but unobtrusively and naturally, the teacher is led to ask the same questions and to indicate the same steps again and again. Thus, in countless problems, we have to ask the question: *What* *is the unknown?* We may vary the words, and ask the same thing in many different ways: What is required? What do you want to find? What are you supposed to seek? The aim of these questions is to focus the student's attention upon the unknown. Sometimes, we obtain the same effect more naturally with a suggestion: *Look at the unknown!* Question and suggestion aim at the same effect; they tend to provoke the same mental operation.

It seemed to the author that it might be worth while to collect and to group questions and suggestions which are typically helpful in discussing problems with students. The list we study contains questions and suggestions of this sort, carefully chosen and arranged; they are equally useful to the problem-solver who works by himself. If the reader is sufficiently acquainted with the list and can see, behind the suggestion, the action suggested, he may realize that the list enumerates, indirectly, *mental operations typically useful for the solution of problems.* These operations are listed in the order in which they are most likely to occur.

3. **Generality** is an important characteristic of the questions and suggestions contained in our list. Take the questions: *What is the unknown? What are the data? What is the condition?* These questions are generally applicable, we can ask them with good effect dealing with all sorts of problems. Their use is not restricted to any subject-matter. Our problem may be algebraic or geometric, mathematical or nonmathematical, theoretical or practical, a serious problem or a mere puzzle; it makes no difference, the questions make sense and might help us to solve the problem.

There is a **restriction,** in fact, but it has nothing to do with the subject-matter. Certain questions and suggestions of the list are applicable to "problems to find" only, not to "problems to prove." If we have a problem of the latter kind we must use different questions; see PROBLEMS TO FIND, PROBLEMS TO PROVE.

4. Common sense. The questions and suggestions of our list are general, but, except for their generality, they are natural, simple, obvious, and proceed from plain common sense. Take the suggestion: *Look at the unknown! And try to think of a familiar problem having the same or a similar unknown.* This suggestion advises you to do what you would do anyhow, without any advice, if you were seriously concerned with your problem. Are you hungry? You wish to obtain food and you think of familiar ways of obtaining food. Have you a problem of geometric construction? You wish to construct a triangle and you think of familiar ways of constructing a triangle. Have you a problem of any kind? You wish to find a certain unknown, and you think of familiar ways of finding such an unknown, or some similar unknown. If you do so you follow exactly the suggestion we quoted from our list. And you are on the right track, too; the suggestion is a good one, it suggests to you a procedure which is very frequently successful.

All the questions and suggestions of our list are natural, simple, obvious, just plain common sense; but they state plain common sense in general terms. They suggest a certain conduct which comes naturally to any person who is seriously concerned with his problem and has some common sense. But the person who behaves the right way usually does not care to express his behavior in clear words and, possibly, he cannot express it so; our list tries to express it so.

5. Teacher and student. Imitation and practice. There are two aims which the teacher may have in view when addressing to his students a question or a suggestion of the list: First, to help the student to solve the problem at hand. Second, to develop the student's ability so that he may solve future problems by himself.

Experience shows that the questions and suggestions of our list, appropriately used, very frequently help the student. They have two common characteristics, common sense and generality. As they proceed from plain common sense they very often come naturally; they could have occurred to the student himself. As they are general, they help unobtrusively; they just indicate a general direction and leave plenty for the student to do.

But the two aims we mentioned before are closely connected; if the student succeeds in solving the problem at hand, he adds a little to his ability to solve problems. Then, we should not forget that our questions are general, applicable in many cases. If the same question is repeatedly helpful, the student will scarcely fail to notice it and he will be induced to ask the question by himself in a similar situation. Asking the question repeatedly, he may succeed once in eliciting the right idea. By such a success, he discovers the right way of using the question, and then he has really assimilated it.

The student may absorb a few questions of our list so well that he is finally able to put to himself the right question in the right moment and to perform the corresponding mental operation naturally and vigorously. Such a student has certainly derived the greatest possible profit from our list. What can the teacher do in order to obtain this best possible result?

Solving problems is a practical skill like, let us say, swimming. We acquire any practical skill by imitation and practice. Trying to swim, you imitate what other people do with their hands and feet to keep their heads above water, and, finally, you learn to swim by practicing swimming. Trying to solve problems, you have to observe and to imitate what other people do when solving problems and, finally, you learn to do problems by doing them.

The teacher who wishes to develop his students' ability to do problems must instill some interest for problems into their minds and give them plenty of opportunity for imitation and practice. If the teacher wishes to develop in his students the mental operations which correspond to the questions and suggestions of our list, he puts these questions and suggestions to the students as often as he can do so naturally. Moreover, when the teacher solves a problem before the class, he should dramatize his ideas a little and he should put to himself the same questions which he uses when helping the students. Thanks to such guidance, the student will eventually discover the right use of these questions and suggestions, and doing so he will acquire something that is more important than the knowledge of any particular mathematical fact.

MAIN DIVISIONS, MAIN QUESTIONS

6. Four phases. Trying to find the solution, we may repeatedly change our point of view, our way of looking at the problem. We have to shift our position again and again. Our conception of the problem is likely to be rather incomplete when we start the work; our outlook is different when we have made some progress; it is again different when we have almost obtained the solution.

In order to group conveniently the questions and suggestions of our list, we shall distinguish four phases of the work. First, we have to *understand* the problem; we have to see clearly what is required. Second, we have to see how the various items are connected, how the unknown is linked to the data, in order to obtain the idea of the solution, to make a *plan*. Third, we *carry out* our plan. Fourth, we *look back* at the completed solution, we review and discuss it.

Each of these phases has its importance. It may happen that a student hits upon an exceptionally bright idea and jumping all preparations blurts out with the solution. Such lucky ideas, of course, are most desirable, but something very undesirable and unfortunate may result if the student leaves out any of the four phases without having a good idea. The worst may happen if the student embarks upon computations or constructions without having *understood* the problem. It is generally useless to carry out details without having seen the main connection, or having made a sort of *plan*. Many mistakes can be avoided if, carrying out his plan, the student *checks each step*. Some of the best effects may be lost if the student fails to reexamine and to *reconsider* the completed solution.

7. Understanding the problem. It is foolish to answer a question that you do not understand. It is sad to work for an end that you do not desire. Such foolish and sad things often happen, in and out of school, but the teacher should try to prevent them from happening in his class. The student should understand the problem. But he should not only understand it, he should also desire its solution. If the student is lacking in understanding or in interest, it is not always his fault; the problem should be well chosen, not too difficult and not too easy, natural and interesting, and some time should be allowed for natural and interesting presentation.

First of all, the verbal statement of the problem must be understood. The teacher can check this, up to a certain extent; he asks the student to repeat the statement, and the student should be able to state the problem fluently. The student should also be able to point out the principal parts of the problem, the unknown, the data, the condition. Hence, the teacher can seldom afford to miss the questions: *What is the unknown? What are the data? What is the condition?*

The student should consider the principal parts of the problem attentively, repeatedly, and from various sides. If there is a figure connected with the problem he should *draw a figure* and point out on it the unknown and the data. If it is necessary to give names to these objects he should *introduce suitable notation;* devoting some attention to the appropriate choice of signs, he is obliged to consider the objects for which the signs have to be chosen. There is another question which may be useful in this preparatory stage provided that we do not expect a definitive answer but just a provisional answer, a guess: *Is it possible to satisfy the condition?*

(In the exposition of Part II [p. 33] "Understanding the problem" is subdivided into two stages: "Getting acquainted" and "Working for better understanding.")

8. Example. Let us illustrate some of the points explained in the foregoing section. We take the following simple problem: *Find the diagonal of a rectangular parallelepiped of which the length, the width, and the height are known.*

In order to discuss this problem profitably, the students must be familiar with the theorem of Pythagoras, and with some of its applications in plane geometry, but they may have very little systematic knowledge in solid geometry. The teacher may rely here upon the student's unsophisticated familiarity with spatial relations.

The teacher can make the problem interesting by making it concrete. The classroom is a rectangular parallelepiped whose dimensions could be measured, and can be estimated; the students have to find, to "measure indirectly," the diagonal of the classroom. The teacher points out the length, the width, and the height of the classroom, indicates the diagonal with a gesture, and enlivens his figure, drawn on the blackboard, by referring repeatedly to the classroom.

The dialogue between the teacher and the students may start as follows:

"*What is the unknown?*"

"The length of the diagonal of a parallelepiped."

"*What are the data?*"

"The length, the width, and the height of the parallelepiped."

"*Introduce suitable notation.* Which letter should denote the unknown?"

"x."

"Which letters would you choose for the length, the width, and the height?"

"a, b, c."

"*What is the condition,* linking $a, b, c,$ and x?"

"x is the diagonal of the parallelepiped of which $a, b,$ and c are the length, the width, and the height."

"Is it a reasonable problem? I mean, *is the condition sufficient to determine the unknown?*"

"Yes, it is. If we know $a, b, c,$ we know the parallelepiped. If the parallelepiped is determined, the diagonal is determined."

9. Devising a plan. We have a plan when we know, or know at least in outline, which calculations, computations, or constructions we have to perform in order to obtain the unknown. The way from understanding the problem to conceiving a plan may be long and tortuous. In fact, the main achievement in the solution of a problem is to conceive the idea of a plan. This idea may emerge gradually. Or, after apparently unsuccessful trials and a period of hesitation, it may occur suddenly, in a flash, as a "bright idea." The best that the teacher can do for the student is to procure for him, by unobtrusive help, a bright idea. The questions and suggestions we are going to discuss tend to provoke such an idea.

In order to be able to see the student's position, the teacher should think of his own experience, of his difficulties and successes in solving problems.

We know, of course, that it is hard to have a good idea if we have little knowledge of the subject, and impossible to have it if we have no knowledge. Good ideas are based on past experience and formerly acquired knowledge. Mere remembering is not enough for a good idea, but we cannot have any good idea without recollecting some pertinent facts; materials alone are not enough for constructing a house but we cannot construct a house without collecting the necessary materials. The materials necessary for solving a mathematical problem are certain relevant items of our formerly acquired mathematical knowledge, as formerly solved problems, or formerly proved theorems. Thus, it is often appropriate to start the work with the question: *Do you know a related problem?*

The difficulty is that there are usually too many problems which are somewhat related to our present problem, that is, have some point in common with it. How can we choose the one, or the few, which are really useful? There is a suggestion that puts our finger on an essential common point: *Look at the unknown! And try to think of a familiar problem having the same or a similar unknown.*

If we succeed in recalling a formerly solved problem which is closely related to our present problem, we are lucky. We should try to deserve such luck; we may deserve it by exploiting it. *Here is a problem related to yours and solved before. Could you use it?*

The foregoing questions, well understood and seriously considered, very often help to start the right train of ideas; but they cannot help always, they cannot work magic. If they do not work, we must look around for some other appropriate point of contact, and explore the various aspects of our problem; we have to vary, to transform, to modify the problem. *Could you restate the problem?* Some of the questions of our list hint specific means to vary the problem, as generalization, specialization, use of analogy, dropping a part of the condition, and so on; the details are important but we cannot go into them now. Variation of the problem may lead to some appropriate auxiliary problem: *If you cannot solve the proposed problem try to solve first some related problem.*

Trying to apply various known problems or theorems, considering various modifications, experimenting with various auxiliary problems, we may stray so far from our original problem that we are in danger of losing it altogether. Yet there is a good question that may bring us back to it: *Did you use all the data? Did you use the whole condition?*

10. Example. We return to the example considered in section 8. As we left it, the students just succeeded in understanding the problem and showed some mild interest in it. They could now have some ideas of their own, some initiative. If the teacher, having watched sharply, cannot detect any sign of such initiative he has to resume carefully his dialogue with the students. He must be prepared to repeat with some modification the questions which the students do not answer. He must be prepared to meet often with the disconcerting silence of the students (which will be indicated by dots . . . . .).

"Do you know a related problem?"

. . . . .

"Look at the unknown! Do you know a problem having the same unknown?"

. . . . .

"Well, *what is the unknown?"*

"The diagonal of a parallelepiped."

"Do you know any *problem with the same unknown?*"

"No. We have not had any problem yet about the diagonal of a parallelepiped."

"Do you know any *problem with a similar unknown?*"

. . . . .

"You see, the diagonal is a segment, the segment of a straight line. Did you never solve a problem whose unknown was the length of a line?"

"Of course, we have solved such problems. For instance, to find a side of a right triangle."

"Good! *Here is a problem related to yours and solved before. Could you use it?*"

. . . . .

"You were lucky enough to remember a problem which is related to your present one and which you solved

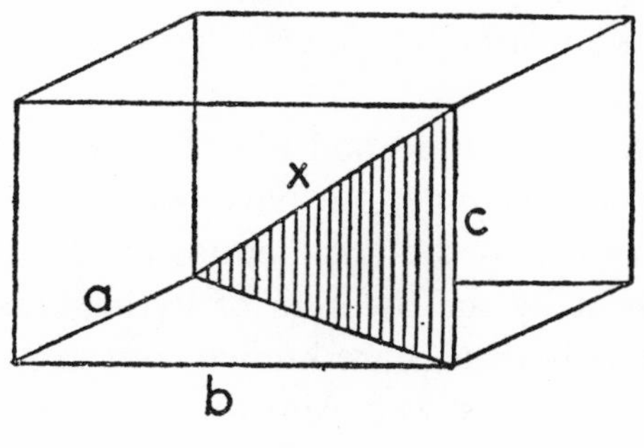

FIG. 1

before. Would you like to use it? *Could you introduce some auxiliary element in order to make its use possible?*"

. . . . .

"Look here, the problem you remembered is about a triangle. Have you any triangle in your figure?"

Let us hope that the last hint was explicit enough to provoke the idea of the solution which is to introduce a right triangle, (emphasized in Fig. 1) of which the required diagonal is the hypotenuse. Yet the teacher should be prepared for the case that even this fairly explicit hint is insufficient to shake the torpor of the students; and so he should be prepared to use a whole gamut of more and more explicit hints.

"Would you like to have a triangle in the figure?"

"What sort of triangle would you like to have in the figure?"

"You cannot find yet the diagonal; but you said that you could find the side of a triangle. Now, what will you do?"

"Could you find the diagonal, if it were a side of a triangle?"

When, eventually, with more or less help, the students succeed in introducing the decisive auxiliary element, the right triangle emphasized in Fig. 1, the teacher should convince himself that the students see sufficiently far ahead before encouraging them to go into actual calculations.

"I think that it was a good idea to draw that triangle. You have now a triangle; but have you the unknown?"

"The unknown is the hypotenuse of the triangle; we can calculate it by the theorem of Pythagoras."

"You can, if both legs are known; but are they?"

"One leg is given, it is c. And the other, I think, is not difficult to find. Yes, the other leg is the hypotenuse of another right triangle."

"Very good! Now I see that you have a plan."

11. Carrying out the plan. To devise a plan, to conceive the idea of the solution is not easy. It takes so much to succeed; formerly acquired knowledge, good mental habits, concentration upon the purpose, and one more thing: good luck. To carry out the plan is much easier; what we need is mainly patience.

The plan gives a general outline; we have to convince

ourselves that the details fit into the outline, and so we have to examine the details one after the other, patiently, till everything is perfectly clear, and no obscure corner remains in which an error could be hidden.

If the student has really conceived a plan, the teacher has now a relatively peaceful time. The main danger is that the student forgets his plan. This may easily happen if the student received his plan from outside, and accepted it on the authority of the teacher; but if he worked for it himself, even with some help, and conceived the final idea with satisfaction, he will not lose this idea easily. Yet the teacher must insist that the student should *check each step.*

We may convince ourselves of the correctness of a step in our reasoning either "intuitively" or "formally." We may concentrate upon the point in question till we see it so clearly and distinctly that we have no doubt that the step is correct; or we may derive the point in question according to formal rules. (The difference between "insight" and "formal proof" is clear enough in many important cases; we may leave further discussion to philosophers.)

The main point is that the student should be honestly convinced of the correctness of each step. In certain cases, the teacher may emphasize the difference between "seeing" and "proving": *Can you see clearly that the step is correct?* But can you also *prove that the step is correct?*

12. Example. Let us resume our work at the point where we left it at the end of section **10**. The student, at last, has got the idea of the solution. He sees the right triangle of which the unknown x is the hypotenuse and the given height c is one of the legs; the other leg is the diagonal of a face. The student must, possibly, be urged to introduce suitable notation. He should choose y to denote that other leg, the diagonal of the face whose sides are a and b. Thus, he may see more clearly the idea of the solution which is to introduce an auxiliary problem whose unknown is y. Finally, working at one right triangle after the other, he may obtain (see Fig. 1)

$$x^2 = y^2 + c^2$$
$$y^2 = a^2 + b^2$$

and hence, eliminating the auxiliary unknown y,

$$x^2 = a^2 + b^2 + c^2$$
$$x = \sqrt{a^2 + b^2 + c^2}\,.$$

The teacher has no reason to interrupt the student if he carries out these details correctly except, possibly, to warn him that he should *check each step*. Thus, the teacher may ask:

"Can you *see clearly* that the triangle with sides x, y, c is a right triangle?"

To this question the student may answer honestly "Yes" but he could be much embarrassed if the teacher, not satisfied with the intuitive conviction of the student, should go on asking:

"But can you *prove* that this triangle is a right triangle?"

Thus, the teacher should rather suppress this question unless the class has had a good initiation in solid geometry. Even in the latter case, there is some danger that the answer to an incidental question may become the main difficulty for the majority of the students.

13. Looking back. Even fairly good students, when they have obtained the solution of the problem and written down neatly the argument, shut their books and look for something else. Doing so, they miss an important and instructive phase of the work. By looking back at the completed solution, by reconsidering and reexamining the result and the path that led to it, they could consolidate their knowledge and develop their ability to solve problems. A good teacher should understand and impress on his students the view that no problem whatever is completely exhausted. There remains always something to do; with sufficient study and penetration, we could improve any solution, and, in any case, we can always improve our understanding of the solution.

The student has now carried through his plan. He has written down the solution, checking each step. Thus, he should have good reasons to believe that his solution is correct. Nevertheless, errors are always possible, especially if the argument is long and involved. Hence, verifications are desirable. Especially, if there is some rapid and intuitive procedure to test either the result or the argument, it should not be overlooked. *Can you check the result? Can you check the argument?*

In order to convince ourselves of the presence or of the quality of an object, we like to see and to touch it. And as we prefer perception through two different senses, so we prefer conviction by two different proofs: *Can you derive the result differently?* We prefer, of course, a short and intuitive argument to a long and heavy one: *Can you see it at a glance?*

One of the first and foremost duties of the teacher is not to give his students the impression that mathematical problems have little connection with each other, and no connection at all with anything else. We have a natural opportunity to investigate the connections of a problem when looking back at its solution. The students will find looking back at the solution really interesting if they have made an honest effort, and have the consciousness of having done well. Then they are eager to see what else they could accomplish with that effort, and how they could do equally well another time. The teacher should encourage the students to imagine cases in which they could utilize again the procedure used, or apply the result obtained. *Can you use the result, or the method, for some other problem?*

14. Example. In section 12, the students finally obtained the solution: If the three edges of a rectangular parallelogram, issued from the same corner, are a, b, c, the diagonal is

$$\sqrt{a^2 + b^2 + c^2}.$$

Can you check the result? The teacher cannot expect a good answer to this question from inexperienced students. The students, however, should acquire fairly early the experience that problems "in letters" have a great advantage over purely numerical problems; if the problem is given "in letters" its result is accessible to several tests to which a problem "in numbers" is not susceptible at all. Our example, although fairly simple, is sufficient to show this. The teacher can ask several questions about the result which the students may readily answer with "Yes"; but an answer "No" would show a serious flaw in the result.

"*Did you use all the data?* Do all the data a, b, c appear in your formula for the diagonal?"

"Length, width, and height play the same role in our question; our problem is symmetric with respect to a, b, c. Is the expression you obtained for the diagonal symmetric in a, b, c? Does it remain unchanged when a, b, c are interchanged?"

"Our problem is a problem of solid geometry: to find the diagonal of a parallelepiped with given dimensions a, b, c. Our problem is analogous to a problem of plane geometry: to find the diagonal of a rectangle with given dimensions a, b. Is the result of our 'solid' problem analogous to the result of the 'plane' problem?"

"If the height c decreases, and finally vanishes, the

parallelepiped becomes a parallelogram. If you put $c = 0$ in your formula, do you obtain the correct formula for the diagonal of the rectangular parallelogram?"

"If the height c increases, the diagonal increases. Does your formula show this?"

"If all three measures a, b, c of the parallelepiped increase in the same proportion, the diagonal also increases in the same proportion. If, in your formula, you substitute $12a, 12b, 12c$ for a, b, c respectively, the expression of the diagonal, owing to this substitution, should also be multiplied by 12. Is that so?"

"If a, b, c are measured in feet, your formula gives the diagonal measured in feet too; but if you change all measures into inches, the formula should remain correct. Is that so?"

(The two last questions are essentially equivalent; see TEST BY DIMENSION.)

These questions have several good effects. First, an intelligent student cannot help being impressed by the fact that the formula passes so many tests. He was convinced before that the formula is correct because he derived it carefully. But now he is more convinced, and his gain in confidence comes from a different source; it is due to a sort of "experimental evidence." Then, thanks to the foregoing questions, the details of the formula acquire new significance, and are linked up with various facts. The formula has therefore a better chance of being remembered, the knowledge of the student is consolidated. Finally, these questions can be easily transferred to similar problems. After some experience with similar problems, an intelligent student may perceive the underlying general ideas: use of all relevant data, variation of the data, symmetry, analogy. If he gets into the habit of directing his attention to such points, his ability to solve problems may definitely profit.

Can you check the argument? To recheck the argument step by step may be necessary in difficult and important cases. Usually, it is enough to pick out "touchy" points for rechecking. In our case, it may be advisable to discuss retrospectively the question which was less advisable to discuss as the solution was not yet attained: Can you *prove* that the triangle with sides x, y, c is a right triangle? (See the end of section **12.**)

Can you use the result or the method for some other problem? With a little encouragement, and after one or two examples, the students easily find applications which consist essentially in giving some *concrete interpretation* to the abstract mathematical elements of the problem. The teacher himself used such a concrete interpretation as he took the room in which the discussion takes place for the parallelepiped of the problem. A dull student may propose, as application, to calculate the diagonal of the cafeteria instead of the diagonal of the classroom. If the students do not volunteer more imaginative remarks, the teacher himself may put a slightly different problem, for instance: "Being given the length, the width, and the height of a rectangular parallelepiped, find the distance of the center from one of the corners."

The students may use the *result* of the problem they just solved, observing that the distance required is one half of the diagonal they just calculated. Or they may use the *method*, introducing suitable right triangles (the latter alternative is less obvious and somewhat more clumsy in the present case).

After this application, the teacher may discuss the configuration of the four diagonals of the parallelepiped, and the six pyramids of which the six faces are the bases, the center the common vertex, and the semidiagonals the lateral edges. When the geometric imagination of the students is sufficiently enlivened, the teacher should come back to his question: *Can you use the result, or the method, for some other problem?* Now there is a better chance that the students may find some more interesting concrete interpretation, for instance, the following:

"In the center of the flat rectangular top of a building which is 21 yards long and 16 yards wide, a flagpole is to be erected, 8 yards high. To support the pole, we need four equal cables. The cables should start from the same point, 2 yards under the top of the pole, and end at the four corners of the top of the building. How long is each cable?"

The students may use the *method* of the problem they solved in detail introducing a right triangle in a vertical plane, and another one in a horizontal plane. Or they may use the *result*, imagining a rectangular parallelepiped of which the diagonal, x, is one of the four cables and the edges are

$$a = 10.5 \qquad b = 8 \qquad c = 6.$$

By straightforward application of the formula, $x = 14.5$.

For more examples, see CAN YOU USE THE RESULT?

15. Various approaches. Let us still retain, for a while, the problem we considered in the foregoing sections **8, 10, 12, 14.** The main work, the discovery of the plan, was described in section **10.** Let us observe that the teacher could have proceeded differently. Starting from the same point as in section **10,** he could have followed a somewhat different line, asking the following questions:

"Do you know any related problem?"

"Do you know an *analogous* problem?"

"You see, the proposed problem is a problem of solid geometry. Could you think of a simpler analogous problem of plane geometry?"

"You see, the proposed problem is about a figure in space, it is concerned with the diagonal of a rectangular parallelepiped. What might be an analogous problem about a figure in the plane? It should be concerned with —the diagonal—of—a rectangular—"

"Parallelogram."

The students, even if they are very slow and indifferent, and were not able to guess anything before, are obliged finally to contribute at least a minute part of the idea. Besides, if the students are so slow, the teacher should not take up the present problem about the parallelepiped without having discussed before, in order to prepare the students, the analogous problem about the parallelogram. Then, he can go on now as follows:

"Here is a problem related to yours and solved before. Can you use it?"

"Should you introduce some auxiliary element in order to make its use possible?"

Eventually, the teacher may succeed in suggesting to the students the desirable idea. It consists in conceiving the diagonal of the given parallelepiped as the diagonal of a suitable parallelogram which must be introduced into the figure (as intersection of the parallelepiped with a plane passing through two opposite edges). The idea is essentially the same as before (section **10**) but the approach is different. In section **10**, the contact with the available knowledge of the students was established through the unknown; a formerly solved problem was recollected because its unknown was the same as that of the proposed problem. In the present section analogy provides the contact with the idea of the solution.

16. The teacher's method of questioning shown in the foregoing sections **8, 10, 12, 14, 15** is essentially this: Begin with a general question or suggestion of our list, and, if necessary, come down gradually to more specific and concrete questions or suggestions till you reach one which elicits a response in the student's mind. If you have to help the student exploit his idea, start again, if possible, from a general question or suggestion contained in the list, and return again to some more special one if necessary; and so on.

Of course, our list is just a first list of this kind; it seems to be sufficient for the majority of simple cases, but there is no doubt that it could be perfected. It is important, however, that the suggestions from which we start should be simple, natural, and general, and that their list should be short.

The suggestions must be simple and natural because otherwise they cannot be *unobtrusive.*

The suggestions must be general, applicable not only to the present problem but to problems of all sorts, if they are to help develop the *ability* of the student and not just a special technique.

The list must be short in order that the questions may be often repeated, unartificially, and under varying circumstances; thus, there is a chance that they will be eventually assimilated by the student and will contribute to the development of a *mental habit.*

It is necessary to come down gradually to specific suggestions, in order that the student may have as great a *share of the work* as possible.

This method of questioning is not a rigid one; fortunately so, because, in these matters, any rigid, mechanical, pedantical procedure is necessarily bad. Our method admits a certain elasticity and variation, it admits various approaches (section **15**), it can be and should be so applied that questions asked by the teacher *could have occurred to the student himself.*

If a reader wishes to try the method here proposed in his class he should, of course, proceed with caution. He should study carefully the example introduced in section **8**, and the following examples in sections **18, 19, 20.** He should prepare carefully the examples which he intends to discuss, considering also various approaches. He should start with a few trials and find out gradually how he can manage the method, how the students take it, and how much time it takes.

17. Good questions and bad questions. If the method of questioning formulated in the foregoing section is well understood it helps to judge, by comparison, the quality of certain suggestions which may be offered with the intention of helping the students.

Let us go back to the situation as it presented itself at the beginning of section **10** when the question was asked: *Do you know a related problem?* Instead of this, with the best intention to help the students, the question may be offered: *Could you apply the theorem of Pythagoras?*

The intention may be the best, but the question is about the worst. We must realize in what situation it was offered; then we shall see that there is a long sequence of objections against that sort of "help."

(1) If the student is near to the solution, he may understand the suggestion implied by the question; but if he is not, he quite possibly will not see at all the point at which the question is driving. Thus the question fails to help where help is most needed.

(2) If the suggestion is understood, it gives the whole secret away, very little remains for the student to do.

(3) The suggestion is of too special a nature. Even if the student can make use of it in solving the present problem, nothing is learned for future problems. The question is not instructive.

(4) Even if he understands the suggestion, the student can scarcely understand how the teacher came to the idea of putting such a question. And how could he, the student, find such a question by himself? It appears as an unnatural surprise, as a rabbit pulled out of a hat; it is really not instructive.

None of these objections can be raised against the procedure described in section **10**, or against that in section **15**.

MORE EXAMPLES

18. A problem of construction. *Inscribe a square in a given triangle. Two vertices of the square should be on the base of the triangle, the two other vertices of the square on the two other sides of the triangle, one on each.*

"*What is the unknown?*"

"A square."

"*What are the data?*"

"A triangle is given, nothing else."

"*What is the condition?*"

"The four corners of the square should be on the perimeter of the triangle, two corners on the base, one corner on each of the other two sides."

"*Is it possible to satisfy the condition?*"

"I think so. I am not so sure."

"You do not seem to find the problem too easy. *If you cannot solve the proposed problem, try to solve first some related problem.* Could you satisfy a *part of the condition?*"

"What do you mean by a part of the condition?"

"You see, the condition is concerned with all the vertices of the square. How many vertices are there?"

"Four."

"A part of the condition would be concerned with less than four vertices. *Keep only a part of the condition, drop the other part.* What part of the condition is easy to satisfy?"

"It is easy to draw a square with two vertices on the perimeter of the triangle—or even one with three vertices on the perimeter!"

"*Draw a figure!*"

The student draws Fig. 2.

"You *kept only a part of the condition,* and you *dropped the other part. How far is the unknown now determined?*"

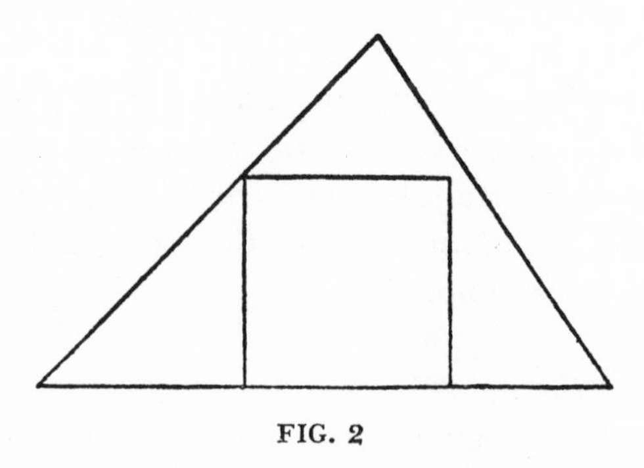

FIG. 2

"The square is not determined if it has only three vertices on the perimeter of the triangle."

"Good! *Draw a figure.*"

The student draws Fig. 3.

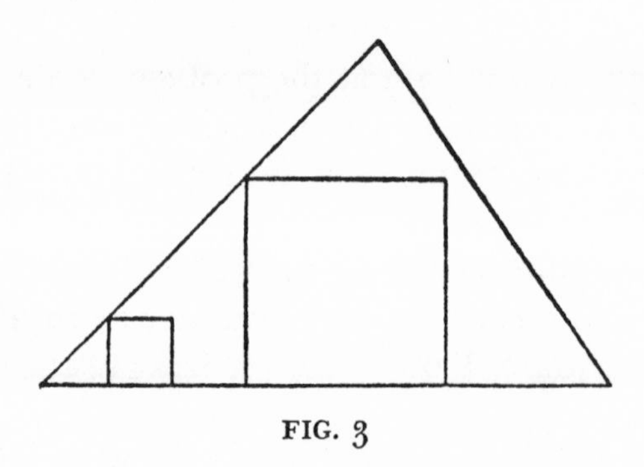

FIG. 3

"The square, as you said, is not determined by the *part of the condition you kept. How can it vary?*"

. . . . .

"Three corners of your square are on the perimeter of the triangle but the fourth corner is not yet there where it should be. Your square, as you said, is undetermined, it can vary; the same is true of its fourth corner. *How can it vary?"*

.....

"Try it experimentally, if you wish. Draw more squares with three corners on the perimeter in the same way as the two squares already in the figure. Draw small squares and large squares. What seems to be the locus of the fourth corner? *How can it vary?"*

The teacher brought the student very near to the idea of the solution. If the student is able to guess that the locus of the fourth corner is a straight line, he has got it.

19. A problem to prove. *Two angles are in different planes but each side of one is parallel to the corresponding side of the other, and has also the same direction. Prove that such angles are equal.*

What we have to prove is a fundamental theorem of solid geometry. The problem may be proposed to students who are familiar with plane geometry and acquainted with those few facts of solid geometry which prepare the present theorem in Euclid's Elements. (The theorem that we have stated and are going to prove is the proposition 10 of Book XI of Euclid.) Not only questions and suggestions quoted from our list are printed in italics but also others which correspond to them as "problems to prove" correspond to "problems to find." (The correspondence is worked out systematically in PROBLEMS TO FIND, PROBLEMS TO PROVE 5, 6.)

"What is the hypothesis?"

"Two angles are in different planes. Each side of one is parallel to the corresponding side of the other, and has also the same direction.

"What is the conclusion?"

"The angles are equal."

"Draw a figure. Introduce suitable notation."

The student draws the lines of Fig. 4 and chooses, helped more or less by the teacher, the letters as in Fig. 4.

"*What is the hypothesis?* Say it, please, using your notation."

"A, B, C are not in the same plane as A', B', C'. And $AB \parallel A'B'$, $AC \parallel A'C'$. Also AB has the same direction as $A'B'$, and AC the same as $A'C'$."

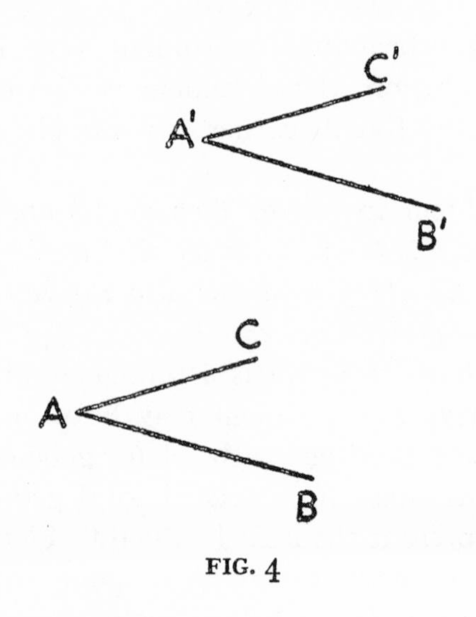

FIG. 4

"*What is the conclusion?*"

"$\angle BAC = \angle B'A'C'$."

"*Look at the conclusion! And try to think of a familiar theorem having the same or a similar conclusion.*"

"If two triangles are congruent, the corresponding angles are equal."

"Very good! Now *here is a theorem related to yours and proved before. Could you use it?*"

"I think so but I do not see yet quite how."

"*Should you introduce some auxiliary element in order to make its use possible?*"

. . . . .

"Well, the theorem which you quoted so well is about

triangles, about a pair of congruent triangles. Have you any triangles in your figure?"

"No. But I could introduce some. Let me join B to C, and B' to C'. Then there are two triangles, $\triangle ABC$, $\triangle A'B'C'$."

"Well done. But what are these triangles good for?"

"To prove the conclusion, $\angle BAC = \angle B'A'C'$."

"Good! If you wish to prove this, what kind of triangles do you need?"

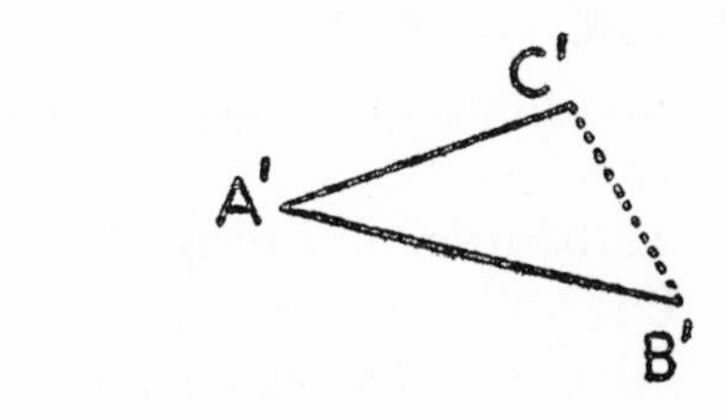

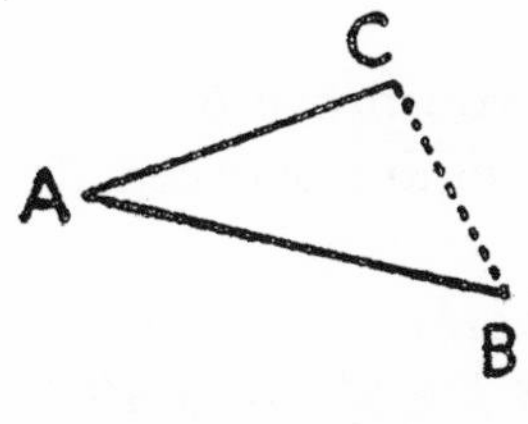

FIG. 5

"Congruent triangles. Yes, of course, I may choose B, C, B', C' so that

$$AB = A'B', AC = A'C'."$$

"Very good! Now, what do you wish to prove?"

"I wish to prove that the triangles are congruent,

$$\triangle ABC = \triangle A'B'C'.$$

If I could prove this, the conclusion $\angle BAC = \angle B'A'C'$ would follow immediately."

"Fine! You have a new aim, you aim at a new conclusion. *Look at the conclusion! And try to think of a familiar theorem having the same or a similar conclusion."*

"Two triangles are congruent if—if the three sides of the one are equal respectively to the three sides of the other."

"Well done. You could have chosen a worse one. Now *here is a theorem related to yours and proved before. Could you use it?"*

"I could use it if I knew that $BC = B'C'$."

"That is right! Thus, what is your aim?"

"To prove that $BC = B'C'$."

"Try to think of a familiar theorem having the same or a similar conclusion."

"Yes, I know a theorem finishing: '. . . then the two lines are equal.' But it does not fit in."

"Should you introduce some auxiliary element in order to make its use possible?"

. . . . .

"You see, how could you prove $BC = B'C'$ when there is no connection in the figure between BC and $B'C'$?"

. . . . .

"Did you use the hypothesis? What is the hypothesis?"

"We suppose that $AB \parallel A'B'$, $AC \parallel A'C'$. Yes, of course, I must use that."

"Did you use the whole hypothesis? You say that $AB \parallel A'B'$. Is that all that you know about these lines?"

"No; AB is also equal to $A'B'$, by construction. They are parallel and equal to each other. And so are AC and $A'C'$."

"Two parallel lines of equal length—it is an interesting configuration. *Have you seen it before?"*

"Of course! Yes! Parallelogram! Let me join A to A', B to B', and C to C'."

"The idea is not so bad. How many parallelograms have you now in your figure?"

"Two. No, three. No, two. I mean, there are two of

which you can prove immediately that they are parallelograms. There is a third which seems to be a parallelogram; I hope I can prove that it is one. And then the proof will be finished!"

We could have gathered from his foregoing answers that the student is intelligent. But after this last remark of his, there is no doubt.

This student is able to guess a mathematical result and to distinguish clearly between proof and guess. He knows also that guesses can be more or less plausible. Really, he did profit something from his mathematics classes; he has some real experience in solving problems, he can conceive and exploit a good idea.

20. A rate problem. *Water is flowing into a conical vessel at the rate* r. *The vessel has the shape of a right circular cone, with horizontal base, the vertex pointing downwards; the radius of the base is* a, *the altitude of the*

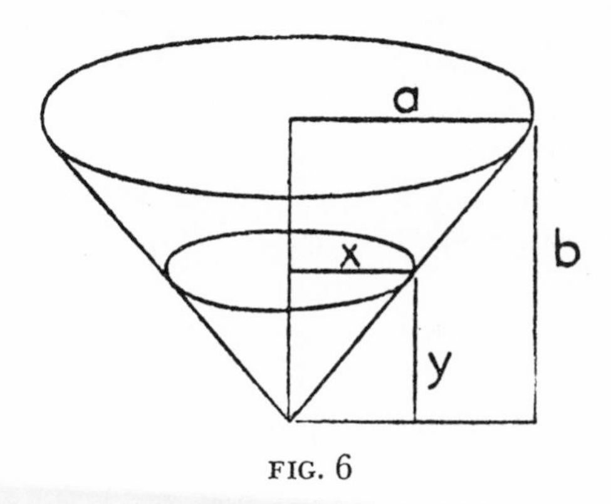

FIG. 6

cone b. *Find the rate at which the surface is rising when the depth of the water is* y. *Finally, obtain the numerical value of the unknown supposing that* a = 4 *ft.*, b = 3 *ft.*, r = 2 *cu. ft. per minute, and* y = 1 *ft.*

The students are supposed to know the simplest rules of differentiation and the notion of "rate of change."

"What are the data?"

"The radius of the base of the cone $a = 4$ ft., the altitude of the cone $b = 3$ ft., the rate at which the water is flowing into the vessel $r = 2$ cu. ft. per minute, and the depth of the water at a certain moment, $y = 1$ ft."

"Correct. The statement of the problem seems to suggest that you should disregard, provisionally, the numerical values, work with the letters, express the unknown in terms of a, b, r, y and only finally, after having obtained the expression of the unknown in letters, substitute the numerical values. I would follow this suggestion. Now, *what is the unknown?*"

"The rate at which the surface is rising when the depth of the water is y."

"What is that? Could you say it in other terms?"

"The rate at which the depth of the water is increasing."

"What is that? *Could you restate it still differently?*"

"The rate of change of the depth of the water."

"That is right, the rate of change of y. But what is the rate of change? *Go back to the definition.*"

"The derivative is the rate of change of a function."

"Correct. Now, is y a function? As we said before, we disregard the numerical value of y. Can you imagine that y changes?"

"Yes, y, the depth of the water, increases as the time goes by."

"Thus, y is a function of what?"

"Of the time t."

"Good. *Introduce suitable notation.* How would you write the 'rate of change of y' in mathematical symbols?"

"$\frac{dy}{dt}$"

"Good. Thus, this is your unknown. You have to express it in terms of a, b, r, y. By the way, one of these data is a 'rate.' Which one?"

"r is the rate at which water is flowing into the vessel."

"What is that? Could you say it in other terms?"

"r is the rate of change of the volume of the water in the vessel."

"What is that? *Could you restate it still differently?* How would you write it in *suitable notation?*"

"$r = \frac{dV}{dt}$."

"What is V?"

"The volume of the water in the vessel at the time t."

"Good. Thus, you have to express $\frac{dy}{dt}$ in terms of a, b, $\frac{dV}{dt}$, y. How will you do it?"

. . . . .

"*If you cannot solve the proposed problem try to solve first some related problem.* If you do not see yet the connection between $\frac{dy}{dt}$ and the data, try to bring in some simpler connection that could serve as a stepping stone."

. . . . .

"Do you not see that there are other connections? For instance, are y and V independent of each other?"

"No. When y increases, V must increase too."

"Thus, there is a connection. What is the connection?"

"Well, V is the volume of a cone of which the altitude is y. But I do not know yet the radius of the base."

"You may consider it, nevertheless. Call it something, say x."

"$V = \frac{\pi x^2 y}{3}$."

"Correct. Now, what about x? Is it independent of y?"

"No. When the depth of the water, y, increases the radius of the free surface, x, increases too."

"Thus, there is a connection. What is the connection?"

"Of course, similar triangles.

$$x : y = a : b."$$

"One more connection, you see. I would not miss profiting from it. Do not forget, you wished to know the connection between V and y."

"I have

$$x = \frac{ay}{b}$$

$$V = \frac{\pi a^2 y^3}{3b^2}."$$

"Very good. This looks like a stepping stone, does it not? But you should not forget your goal. *What is the unknown?*"

"Well, $\frac{dy}{dt}$."

"You have to find a connection between $\frac{dy}{dt}$, $\frac{dV}{dt}$, and other quantities. And here you have one between y, V, and other quantities. What to do?"

"Differentiate! Of course!

$$\frac{dV}{dt} = \frac{\pi a^2 y^2}{b^2} \frac{dy}{dt}.$$

Here it is."

"Fine! And what about the numerical values?"

"If $a = 4$, $b = 3$, $\frac{dV}{dt} = r = 2$, $y = 1$, then

$$2 = \frac{\pi \times 16 \times 1}{9} \frac{dy}{dt}."$$

PART II. HOW TO SOLVE IT
A DIALOGUE

Getting Acquainted

Where should I start? Start from the statement of the problem.

What can I do? Visualize the problem as a whole as clearly and as vividly as you can. Do not concern yourself with details for the moment.

What can I gain by doing so? You should understand the problem, familiarize yourself with it, impress its purpose on your mind. The attention bestowed on the problem may also stimulate your memory and prepare for the recollection of relevant points.

Working for Better Understanding

Where should I start? Start again from the statement of the problem. Start when this statement is so clear to you and so well impressed on your mind that you may lose sight of it for a while without fear of losing it altogether.

What can I do? Isolate the principal parts of your problem. The hypothesis and the conclusion are the principal parts of a "problem to prove"; the unknown, the data, and the conditions are the principal parts of a "problem to find." Go through the principal parts of your problem, consider them one by one, consider them in turn, consider them in various combinations, relating each detail to other details and each to the whole of the problem.

What can I gain by doing so? You should prepare and clarify details which are likely to play a role afterwards.

Hunting for the Helpful Idea

Where should I start? Start from the consideration of the principal parts of your problem. Start when these principal parts are distinctly arranged and clearly conceived, thanks to your previous work, and when your memory seems responsive.

What can I do? Consider your problem from various sides and seek contacts with your formerly acquired knowledge.

Consider your problem from various sides. Emphasize different parts, examine different details, examine the same details repeatedly but in different ways, combine the details differently, approach them from different sides. Try to see some new meaning in each detail, some new interpretation of the whole.

Seek contacts with your formerly acquired knowledge. Try to think of what helped you in similar situations in the past. Try to recognize something familiar in what you examine, try to perceive something useful in what you recognize.

What could I perceive? A helpful idea, perhaps a decisive idea that shows you at a glance the way to the very end.

How can an idea be helpful? It shows you the whole of the way or a part of the way; it suggests to you more or less distinctly how you can proceed. Ideas are more or less complete. You are lucky if you have any idea at all.

What can I do with an incomplete idea? You should consider it. If it looks advantageous you should consider it longer. If it looks reliable you should ascertain how far it leads you, and reconsider the situation. The situation has changed, thanks to your helpful idea. Consider the new situation from various sides and seek contacts with your formerly acquired knowledge.

What can I gain by doing so again? You may be lucky and have another idea. Perhaps your next idea will lead you to the solution right away. Perhaps you need a few more helpful ideas after the next. Perhaps you will be led astray by some of your ideas. Nevertheless you should be grateful for all new ideas, also for the lesser ones, also for the hazy ones, also for the supplementary ideas adding some precision to a hazy one, or attempting the correction of a less fortunate one. Even if you do not have any appreciable new ideas for a while you should be grateful if your conception of the problem becomes more complete or more coherent, more homogeneous or better balanced.

Carrying Out the Plan

Where should I start? Start from the lucky idea that led you to the solution. Start when you feel sure of your grasp of the main connection and you feel confident that you can supply the minor details that may be wanting.

What can I do? Make your grasp quite secure. Carry through in detail all the algebraic or geometric operations which you have recognized previously as feasible. Convince yourself of the correctness of each step by formal reasoning, or by intuitive insight, or both ways if you can. If your problem is very complex you may distinguish "great" steps and "small" steps, each great step being composed of several small ones. Check first the great steps, and get down to the smaller ones afterwards.

What can I gain by doing so? A presentation of the solution each step of which is correct beyond doubt.

Looking Back

Where should I start? From the solution, complete and correct in each detail.

What can I do? Consider the solution from various sides and seek contacts with your formerly acquired knowledge.

Consider the details of the solution and try to make them as simple as you can; survey more extensive parts of the solution and try to make them shorter; try to see the whole solution at a glance. Try to modify to their advantage smaller or larger parts of the solution, try to improve the whole solution, to make it intuitive, to fit it into your formerly acquired knowledge as naturally as possible. Scrutinize the method that led you to the solution, try to see its point, and try to make use of it for other problems. Scrutinize the result and try to make use of it for other problems.

What can I gain by doing so? You may find a new and better solution, you may discover new and interesting facts. In any case, if you get into the habit of surveying and scrutinizing your solutions in this way, you will acquire some knowledge well ordered and ready to use, and you will develop your ability of solving problems.

PART III. SHORT DICTIONARY OF HEURISTIC

Analogy is a sort of similarity. Similar objects agree with each other in some respect, analogous objects *agree in certain relations* of their respective parts.

1. A rectangular parallelogram is analogous to a rectangular parallelepiped. In fact, the relations between the sides of the parallelogram are similar to those between the faces of the parallelepiped:

Each side of the parallelogram is parallel to just one other side, and is perpendicular to the remaining sides.

Each face of the parallelepiped is parallel to just one other face, and is perpendicular to the remaining faces.

Let us agree to call a side a "bounding element" of the parallelogram and a face a "bounding element" of the parallelepiped. Then, we may contract the two foregoing statements into one that applies equally to both figures:

Each bounding element is parallel to just one other bounding element and is perpendicular to the remaining bounding elements.

Thus, we have expressed certain relations which are common to the two systems of objects we compared, sides of the rectangle and faces of the rectangular parallelepiped. The analogy of these systems consists in this community of relations.

2. Analogy pervades all our thinking, our everyday speech and our trivial conclusions as well as artistic ways of expression and the highest scientific achievements. Analogy is used on very different levels. People

often use vague, ambiguous, incomplete, or incompletely clarified analogies, but analogy may reach the level of mathematical precision. All sorts of analogy may play a role in the discovery of the solution and so we should not neglect any sort.

3. We may consider ourselves lucky when, trying to solve a problem, we succeed in discovering a *simpler analogous problem.* In section **15**, our original problem was concerned with the diagonal of a rectangular parallelepiped; the consideration of a simpler analogous problem, concerned with the diagonal of a rectangle, led us to the solution of the original problem. We are going to discuss one more case of the same sort. We have to solve the following problem:

Find the center of gravity of a homogeneous tetrahedron.

Without knowledge of the integral calculus, and with little knowledge of physics, this problem is not easy at all; it was a serious scientific problem in the days of Archimedes or Galileo. Thus, if we wish to solve it with as little preliminary knowledge as possible, we should look around for a simpler analogous problem. The corresponding problem in the plane occurs here naturally:

Find the center of gravity of a homogeneous triangle.

Now, we have two questions instead of one. But two questions may be easier to answer than just one question —provided that the two questions are intelligently connected.

4. Laying aside, for the moment, our original problem concerning the tetrahedron, we concentrate upon the simpler analogous problem concerning the triangle. To solve this problem, we have to know something about centers of gravity. The following principle is plausible and presents itself naturally.

If a system of masses S *consists of parts, each of which has its center of gravity in the same plane, then this plane contains also the center of gravity of the whole system* S.

This principle yields all that we need in the case of the triangle. First, it implies that the center of gravity of the triangle lies in the plane of the triangle. Then, we may consider the triangle as consisting of fibers (thin strips, "infinitely narrow" parallelograms) parallel to a certain side of the triangle (the side AB in Fig. 7). The center of gravity of each fiber (of any parallelogram) is, obviously, its midpoint, and all these midpoints lie on the line joining the vertex C opposite to the side AB to the midpoint M of AB (see Fig. 7).

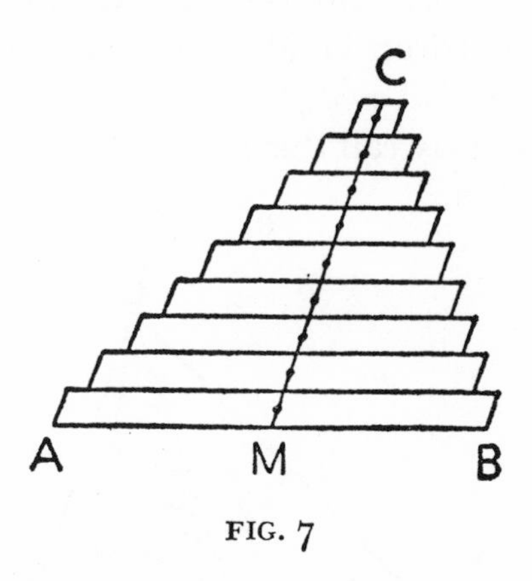

FIG. 7

Any plane passing through the median CM of the triangle contains the centers of gravity of all parallel fibers which constitute the triangle. Thus, we are led to the conclusion that the center of gravity of the whole triangle lies on the same median. Yet it must lie on the other two medians just as well, it must be the *common point of intersection of all three medians.*

It is desirable to verify now by pure geometry, independently of any mechanical assumption, that the three medians meet in the same point.

5. After the case of the triangle, the case of the tetrahedron is fairly easy. We have now solved a problem analogous to our proposed problem and, having solved it, we have a *model to follow*.

In solving the analogous problem which we use now as a model, we conceived the triangle *ABC* as consisting of fibers parallel to one of its sides, *AB*. Now, we conceive the tetrahedron *ABCD* as consisting of fibers parallel to one of its edges, *AB*.

The midpoints of the fibers which constitute the triangle lie all on the same straight line, a median of the triangle, joining the midpoint *M* of the side *AB* to the opposite vertex *C*. The midpoints of the fibers which constitute the tetrahedron lie all in the same plane, joining the midpoint *M* of the edge *AB* to the opposite edge *CD* (see Fig. 8); we may call this plane *MCD* a *median plane* of the tetrahedron.

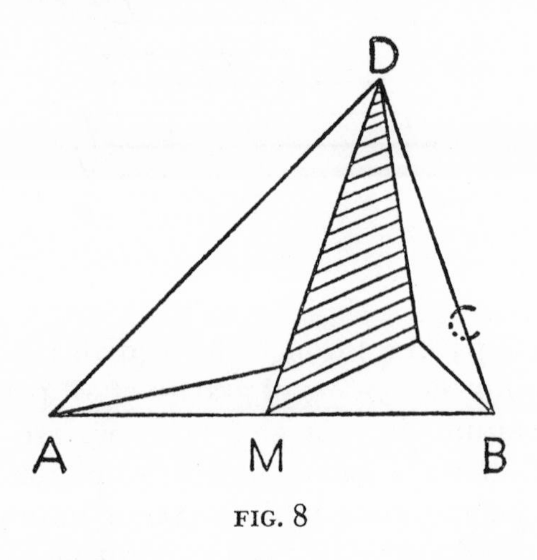

FIG. 8

In the case of the triangle, we had three medians like *MC*, each of which has to contain the center of gravity of the triangle. Therefore, these three medians must meet in one point which is precisely the center of gravity. In the case of the tetrahedron we have six median planes like *MCD*, joining the midpoint of some edge to the opposite edge, each of which has to contain the center of gravity of the tetrahedron. Therefore, these six median planes must meet in one point which is precisely the center of gravity.

6. Thus, we have solved the problem of the center of gravity of the homogeneous tetrahedron. To complete our solution, it is desirable to verify now by pure geometry, independently of mechanical considerations, that the six median planes mentioned pass through the same point.

When we had solved the problem of the center of gravity of the homogeneous triangle, we found it desirable to verify, in order to complete our solution, that the three medians of the triangle pass through the same point. This problem is analogous to the foregoing but visibly simpler.

Again we may use, in solving the problem concerning the tetrahedron, the simpler analogous problem concerning the triangle (which we may suppose here as solved). In fact, consider the three median planes, passing through the three edges *DA, DB, DC* issued from the vertex *D*; each passes also through the midpoint of the opposite edge (the median plane through *DC* passes through *M*, see Fig. 8). Now, these three median planes intersect the plane of △ *ABC* in the three medians of this triangle. These three medians pass through the same point (this is the result of the simpler analogous problem) and this point, just as *D*, is a common point of the three median planes. The straight line, joining the two common points, is common to all three median planes.

We proved that those 3 among the 6 median planes which pass through the vertex *D* have a common straight line. The same must be true of those 3 median planes which pass through A; and also of the 3 median planes through B; and also of the 3 through C. Connecting these facts suitably, we may prove that the 6 median planes have a common point. (The 3 median planes passing through the sides of $\triangle ABC$ determine a common point, and 3 lines of intersection which meet in the common point. Now, by what we have just proved, through each line of intersection one more median plane must pass.)

7. Both under 5 and under 6 we used a simpler analogous problem, concerning the triangle, to solve a problem about the tetrahedron. Yet the two cases are different in an important respect. Under 5, we used the *method* of the simpler analogous problem whose solution we imitated point by point. Under 6, we used the *result* of the simpler analogous problem, and we did not care how this result had been obtained. Sometimes, we may be able to use *both the method and the result* of the simpler analogous problem. Even our foregoing example shows this if we regard the considerations under 5 and 6 as different parts of the solution of the same problem.

Our example is typical. In solving a proposed problem, we can often use the solution of a simpler analogous problem; we may be able to use its method, or its result, or both. Of course, in more difficult cases, complications may arise which are not yet shown by our example. Especially, it can happen that the solution of the analogous problem cannot be immediately used for our original problem. Then, it may be worth while to reconsider the solution, to vary and to modify it till, after having tried various forms of the solution, we find eventually one that can be extended to our original problem.

8. It is desirable to foresee the result, or, at least, some features of the result, with some degree of plausibility. Such plausible forecasts are often based on analogy.

Thus, we may know that the center of gravity of a homogeneous triangle coincides with the center of gravity of its three vertices (that is, of three material points with equal masses, placed in the vertices of the triangle). Knowing this, we may conjecture that the center of gravity of a homogeneous tetrahedron coincides with the center of gravity of its four vertices.

This conjecture is an "inference by analogy." Knowing that the triangle and the tetrahedron are alike in many respects, we conjecture that they are alike in one more respect. It would be foolish to regard the plausibility of such conjectures as certainty, but it would be just as foolish, or even more foolish, to disregard such plausible conjectures.

Inference by analogy appears to be the most common kind of conclusion, and it is possibly the most essential kind. It yields more or less plausible conjectures which may or may not be confirmed by experience and stricter reasoning. The chemist, experimenting on animals in order to foresee the influence of his drugs on humans, draws conclusions by analogy. But so did a small boy I knew. His pet dog had to be taken to the veterinary, and he inquired:

"Who is the veterinary?"

"The animal doctor."

"Which animal is the animal doctor?"

9. An analogical conclusion from many parallel cases is stronger than one from fewer cases. Yet quality is still more important here than quantity. Clear-cut analogies weigh more heavily than vague similarities, systematically arranged instances count for more than random collections of cases.

In the foregoing (under 8) we put forward a conjecture about the center of gravity of the tetrahedron. This conjecture was supported by analogy; the case of the tetrahedron is analogous to that of the triangle. We may strengthen the conjecture by examining one more analogous case, the case of a homogeneous rod (that is, a straight line-segment of uniform density).

The analogy between

segment triangle tetrahedron

has many aspects. A segment is contained in a straight line, a triangle in a plane, a tetrahedron in space. Straight line-segments are the simplest one-dimensional bounded figures, triangles the simplest polygons, tetrahedrons the simplest polyhedrons.

The segment has 2 zero-dimensional bounding elements (2 end-points) and its interior is one-dimensional.

The triangle has 3 zero-dimensional and 3 one-dimensional bounding elements (3 vertices, 3 sides) and its interior is two-dimensional.

The tetrahedron has 4 zero-dimensional, 6 one-dimensional, and 4 two-dimensional bounding elements (4 vertices, 6 edges, 4 faces), and its interior is three-dimensional.

These numbers can be assembled into a table. The successive columns contain the numbers for the zero-, one-, two-, and three-dimensional elements, the successive rows the numbers for the segment, triangle, and tetrahedron:

2	1		
3	3	1	
4	6	4	1

Very little familiarity with the powers of a binomial is needed to recognize in these numbers a section of Pascal's triangle. We found a remarkable regularity in segment, triangle, and tetrahedron.

10. If we have experienced that the objects we compare are closely connected, "inferences by analogy," as the following, may have a certain weight with us.

The center of gravity of a homogeneous rod coincides with the center of gravity of its 2 end-points. The center of gravity of a homogeneous triangle coincides with the center of gravity of its 3 vertices. Should we not suspect that the center of gravity of a homogeneous tetrahedron coincides with the center of gravity of its 4 vertices?

Again, the center of gravity of a homogeneous rod divides the distance between its end-points in the proportion 1 : 1. The center of gravity of a triangle divides the distance between any vertex and the midpoint of the opposite side in the proportion 2 : 1. Should we not suspect that the center of gravity of a homogeneous tetrahedron divides the distance between any vertex and the center of gravity of the opposite face in the proportion 3 : 1?

It appears extremely unlikely that the conjectures suggested by these questions should be wrong, that such a beautiful regularity should be spoiled. The feeling that harmonious simple order cannot be deceitful guides the discoverer both in the mathematical and in the other sciences, and is expressed by the Latin saying: *simplex sigillum veri* (simplicity is the seal of truth).

[The preceding suggests an extension to n dimensions. It appears unlikely that what is true in the first three dimensions, for $n = 1, 2, 3$, should cease to be true for higher values of n. This conjecture is an "inference by induction"; it illustrates that induction is naturally based on analogy. See INDUCTION AND MATHEMATICAL INDUCTION.]

[11. We finish the present section by considering briefly the most important cases in which analogy attains the precision of mathematical ideas.

(I) Two systems of mathematical objects, say S and S', are so connected that certain relations between the objects of S are governed by the same laws as those between the objects of S'.

This kind of analogy between S and S' is exemplified by what we have discussed under 1; take as S the sides of a rectangle, as S' the faces of a rectangular parallelepiped.

(II) There is a one-one correspondence between the objects of the two systems S and S', preserving certain relations. That is, if such a relation holds between the objects of one system, the same relation holds between the corresponding objects of the other system. Such a connection between two systems is a very precise sort of analogy; it is called isomorphism (or holohedral isomorphism).

(III) There is a one-many correspondence between the objects of the two systems S and S' preserving certain relations. Such a connection (which is important in various branches of advanced mathematical study, especially in the Theory of Groups, and need not be discussed here in detail) is called merohedral isomorphism (or homomorphism; homoiomorphism would be, perhaps, a better term). Merohedral isomorphism may be considered as another very precise sort of analogy.]

Auxiliary elements. There is much more in our conception of the problem at the end of our work than was in it as we started working (PROGRESS AND ACHIEVEMENT, 1). As our work progresses, we add new elements to those originally considered. An element that we introduce in the hope that it will further the solution is called an *auxiliary element.*

1. There are various kinds of auxiliary elements. Solving a geometric problem, we may introduce new lines into our figure, *auxiliary lines.* Solving an algebraic problem, we may introduce an *auxiliary unknown* (AUXILIARY PROBLEMS, 1). An *auxiliary theorem* is a theorem whose proof we undertake in the hope of promoting the solution of our original problem.

2. There are various reasons for introducing auxiliary elements. We are glad when we have succeeded in recollecting a *problem related to ours and solved before.* It is probable that we can use such a problem but we do not know yet how to use it. For instance, the problem which we are trying to solve is a geometric problem, and the related problem which we have solved before and have now succeeded in recollecting is a problem about triangles. Yet there is no triangle in our figure; in order to make any use of the problem recollected we must have a triangle; therefore, we have to introduce one, by adding suitable auxiliary lines to our figure. In general, having recollected a formerly solved related problem and wishing to use it for our present one, we must often ask: *Should we introduce some auxiliary element in order to make its use possible?* (The example in section **10** is typical.)

Going back to definitions, we have another opportunity to introduce auxiliary elements. For instance, explicating the definition of a circle we should not only mention its center and its radius, but we should also introduce these geometric elements into our figure. Without introducing them, we could not make any concrete use of the definition; stating the definition without drawing something is mere lip-service.

Trying to use known results and going back to definitions are among the best reasons for introducing auxiliary elements; but they are not the only ones. We may add auxiliary elements to the conception of our problem in order to make it fuller, more suggestive, more familiar although we scarcely know yet explicitly how we shall be able to use the elements added. We may just feel that it is a "bright idea" to conceive the problem that way with such and such elements added.

We may have this or that reason for introducing an

auxiliary element, but we should have some reason. We should not introduce auxiliary elements wantonly.

3. *Example.* Construct a triangle, being given one angle, the altitude drawn from the vertex of the given angle, and the perimeter of the triangle.

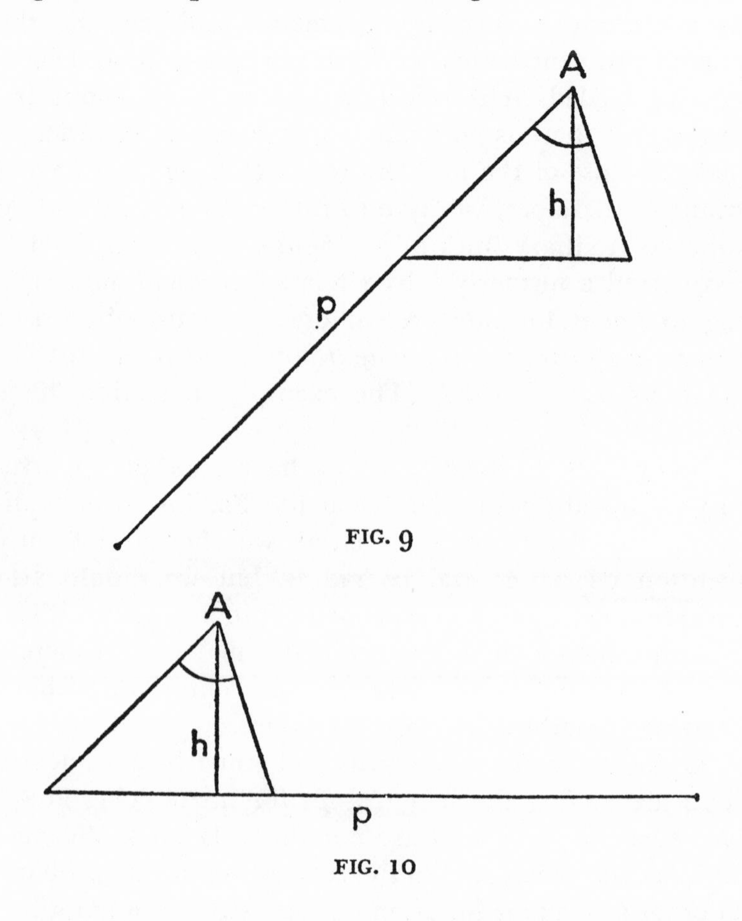

FIG. 9

FIG. 10

We *introduce suitable notation.* Let α denote the given angle, h the given altitude drawn from the vertex A of α and p the given perimeter. We *draw a figure* in which we easily place α and h. *Have we used all the data?* No, our figure does not contain the given length p, equal to the perimeter of the triangle. Therefore we must introduce p. But how?

We may attempt to introduce p in various ways. The attempts exhibited in Figs. 9, 10 appear clumsy. If we try to make clear to ourselves why they appear so unsatisfactory, we may perceive that it is for lack of symmetry.

In fact, the triangle has three unknown sides a, b, c. We call a, as usual, the side opposite to A; we know that

$$a + b + c = p.$$

Now, the sides b and c play the same role; they are interchangeable; our problem is symmetric with respect to b and c. But b and c do not play the same role in our figures 9, 10; placing the length p we treated b and c differently; the figures 9 and 10 spoil the natural symmetry of the problem with respect to b and c. We should place p so that it has the same relation to b as to c.

This consideration may be helpful in suggesting to place the length p as in Fig. 11. We add to the side a of

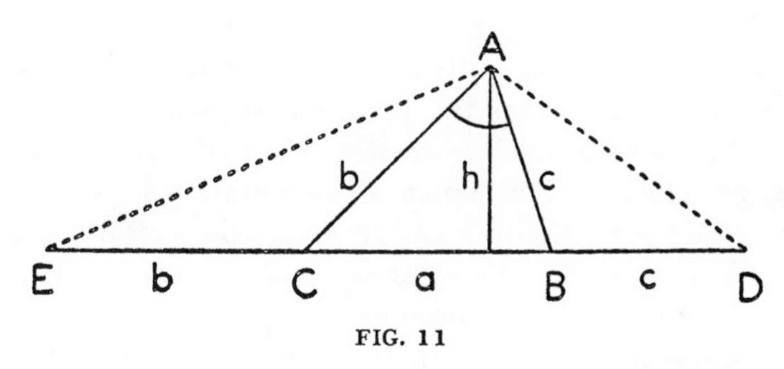

FIG. 11

the triangle the segment CE of length b on one side and the segment BD of the length c on the other side so that p appears in Fig. 11 as the line ED of length

$$b + a + c = p.$$

If we have some little experience in solving problems of construction, we shall not fail to introduce into the figure, along with *ED*, the auxiliary lines *AD* and *AE*, each of which is the base of an isosceles triangle. In fact, it is not unreasonable to introduce elements into the problem which are particularly simple and familiar, as isosceles triangle.

We have been quite lucky in introducing our auxiliary lines. Examining the new figure we may discover that $\angle EAD$ has a simple relation to the given angle α. In fact, we find using the isosceles triangles $\triangle ABD$ and $\triangle ACE$ that $\angle DAE = \frac{\alpha}{2} + 90°$. After this remark, it is natural to try the construction of $\triangle DAE$. Trying this construction, we introduce an auxiliary problem which is much easier than the original problem.

4. Teachers and authors of textbooks should not forget that the intelligent student and THE INTELLIGENT READER are not satisfied by verifying that the steps of a reasoning are correct but also want to know the motive and the purpose of the various steps. The introduction of an auxiliary element is a conspicuous step. If a tricky auxiliary line appears abruptly in the figure, without any motivation, and solves the problem surprisingly, intelligent students and readers are disappointed; they feel that they are cheated. Mathematics is interesting in so far as it occupies our reasoning and inventive powers. But there is nothing to learn about reasoning and invention if the motive and purpose of the most conspicuous step remain incomprehensible. To make such steps comprehensible by suitable remarks (as in the foregoing, under 3) or by carefully chosen questions and suggestions (as in sections **10, 18, 19, 20**) takes a lot of time and effort; but it may be worth while.

Auxiliary problem is a problem which we consider, not for its own sake, but because we hope that its consideration may help us to solve another problem, our original problem. The original problem is the end we wish to attain, the auxiliary problem a means by which we try to attain our end.

An insect tries to escape through the windowpane, tries the same again and again, and does not try the next window which is open and through which it came into the room. A man is able, or at least should be able, to act more intelligently. Human superiority consists in going around an obstacle that cannot be overcome directly, in devising a suitable auxiliary problem when the original problem appears insoluble. To devise an auxiliary problem is an important operation of the mind. To raise a clear-cut new problem subservient to another problem, to conceive distinctly as an end what is means to another end, is a refined achievement of the intelligence. It is an important task to learn (or to teach) how to handle auxiliary problems intelligently.

1. *Example.* Find x, satisfying the equation

$$x^4 - 13x^2 + 36 = 0.$$

If we observe that $x^4 = (x^2)^2$ we may see the advantage of introducing

$$y = x^2.$$

We obtain now a new problem: Find y, satisfying the equation

$$y^2 - 13y + 36 = 0.$$

The new problem is an auxiliary problem; we intend to use it as a means of solving our original problem. The unknown of our auxiliary problem, y, is appropriately called *auxiliary unknown.*

2. *Example.* Find the diagonal of a rectangular parallelepiped being given the lengths of three edges drawn from the same corner.

Trying to solve this problem (section 8) we may be led, by analogy (section 15), to another problem: Find the diagonal of a rectangular parallelogram being given the lengths of two sides drawn from the same vertex.

The new problem is an auxiliary problem; we consider it because we hope to derive some profit for the original problem from its consideration.

3. *Profit.* The profit that we derive from the consideration of an auxiliary problem may be of various kinds. We may use the *result* of the auxiliary problem. Thus, in example 1, having found by solving the quadratic equation for y that y is equal to 4 or to 9, we infer that $x^2 = 4$ or $x^2 = 9$ and derive hence all possible values of x. In other cases, we may use the *method* of the auxiliary problem. Thus, in example 2, the auxiliary problem is a problem of plane geometry; it is analogous to, but simpler than, the original problem which is a problem of solid geometry. It is reasonable to introduce an auxiliary problem of this kind in the hope that it will be instructive, that it will give us opportunity to familiarize ourselves with certain methods, operations, or tools, which we may use afterwards for our original problem. In example 2, the choice of the auxiliary problem is rather lucky; examining it closely we find that we can use both its method and its result. (See section 15, and DID YOU USE ALL THE DATA?)

4. *Risk.* We take away from the original problem the time and the effort that we devote to the auxiliary problem. If our investigation of the auxiliary problem fails, the time and effort we devoted to it may be lost. Therefore, we should exercise our judgment in choosing an auxiliary problem. We may have various good reasons for our choice. The auxiliary problem may appear more accessible than the original problem; or it may appear

instructive; or it may have some sort of aesthetic appeal. Sometimes the only advantage of the auxiliary problem is that it is new and offers unexplored possibilities; we choose it because we are tired of the original problem all approaches to which seem to be exhausted.

5. *How to find one.* The discovery of the solution of the proposed problem often depends on the discovery of a suitable auxiliary problem. Unhappily, there is no infallible method of discovering suitable auxiliary problems as there is no infallible method of discovering the solution. There are, however, questions and suggestions which are frequently helpful, as LOOK AT THE UNKNOWN. We are often led to useful auxiliary problems by VARIATION OF THE PROBLEM.

6. *Equivalent problems.* Two problems are *equivalent* if the solution of each involves the solution of the other. Thus, in our example 1, the original problem and the auxiliary problem are equivalent.

Consider the following theorems:

A. In any equilateral triangle, each angle is equal to 60°.

B. In any equiangular triangle, each angle is equal to 60°.

These two theorems are not identical. They contain different notions; one is concerned with equality of the sides, the other with equality of the angles of a triangle. But each theorem follows from the other. Therefore, the problem to prove A is equivalent to the problem to prove B.

If we are required to prove A, there is a certain advantage in introducing, as an auxiliary problem, the problem to prove B. The theorem B is a little easier to prove than A and, what is more important, we may *foresee* that B is easier than A, we may judge so, we may find plausible from the outset that B is easier than A. In fact, the theorem B, concerned only with angles, is more "homogeneous" than the theorem A which is concerned with both angles and sides.

The passage from the original problem to the auxiliary problem is called *convertible* reduction, or *bilateral* reduction, or *equivalent* reduction if these two problems, the original and the auxiliary, are equivalent. Thus, the reduction of A to B (see above) is convertible and so is the reduction in example 1. Convertible reductions are, in a certain respect, more important and more desirable than other ways to introduce auxiliary problems, but auxiliary problems which are not equivalent to the original problem may also be very useful; take example 2.

7. *Chains of equivalent auxiliary problems* are frequent in mathematical reasoning. We are required to solve a problem A; we cannot see the solution, but we may find that A is equivalent to another problem B. Considering B we may run into a third problem C equivalent to B. Proceeding in the same way, we reduce C to D, and so on, until we come upon a last problem L whose solution is known or immediate. Each problem being equivalent to the preceding, the last problem L must be equivalent to our original problem A. Thus we are able to infer the solution of the original problem A from the problem L which we attained as the last link in a chain of auxiliary problems.

Chains of problems of this kind were noticed by the Greek mathematicians as we may see from an important passage of PAPPUS. For an illustration, let us reconsider our example 1. Let us call (A) the condition imposed upon the unknown x:

$$\text{(A)} \qquad x^4 - 13x^2 + 36 = 0.$$

One way of solving the problem is to transform the proposed condition into another condition which we shall call (B):

$$\text{(B)} \qquad (2x^2)^2 - 2(2x^2)13 + 144 = 0.$$

Observe that the conditions (A) and (B) are different. They are only slightly different if you wish to say so, they are certainly equivalent as you may easily convince yourself, but they are definitely not identical. The passage from (A) to (B) is not only correct but has a clear-cut purpose, obvious to anybody who is familiar with the solution of quadratic equations. Working further in the same direction we transform the condition (B) into still another condition (C):

$$\text{(C)} \qquad (2x^2)^2 - 2(2x^2)13 + 169 = 25.$$

Proceeding in the same way, we obtain

$$\text{(D)} \qquad (2x^2 - 13)^2 = 25$$

$$\text{(E)} \qquad 2x^2 - 13 = \pm 5$$

$$\text{(F)} \qquad x^2 = \frac{13 \pm 5}{2}$$

$$\text{(G)} \qquad x = \pm\sqrt{\frac{13 \pm 5}{2}}$$

$$\text{(H)} \qquad x = 3, \text{ or } -3, \text{ or } 2, \text{ or } -2.$$

Each reduction that we made was convertible. Thus, the last condition (H) is equivalent to the first condition (A) so that $3, -3, 2, -2$ are all possible solutions of our original equation.

In the foregoing, we derived from an original condition (A) a sequence of conditions (B), (C), (D), . . . each of which was equivalent to the foregoing. This point deserves the greatest care. Equivalent conditions are satisfied by the same objects. Therefore, if we pass from a proposed condition to a new condition equivalent to it, we have the same solutions. But if we pass from a proposed condition to a narrower one, we lose solutions, and if we pass to a wider one we admit improper, adventitious solutions which have nothing to do with the proposed problem. If, in a series of successive reductions, we pass to a narrower and then again to a wider condition we may lose track of the original problem completely. In order to avoid this danger, we must check carefully the nature of each newly introduced condition: Is it equivalent to the original condition? This question is still more important when we do not deal with a single equation as here but with a system of equations, or when the condition is not expressed by equations as, for instance, in problems of geometric construction.

(Compare PAPPUS, especially comments 2, 3, 4, 8. The description on p. 143, lines 4–21, is unnecessarily restricted; it describes a chain of "problems to find," each of which has a different unknown. The example considered here has just the opposite speciality: all problems of the chain have the same unknown and differ only in the form of the condition. Of course, no such restriction is necessary.)

8. *Unilateral reduction.* We have two problems, A and B, both unsolved. If we could solve A we could hence derive the full solution of B. But not conversely; if we could solve B, we would obtain, possibly, some information about A, but we would not know how to derive the full solution of A from that of B. In such a case, more is achieved by the solution of A than by the solution of B. Let us call A the *more ambitious,* and B the *less ambitious* of the two problems.

If, from a proposed problem, we pass either to a more ambitious or to a less ambitious auxiliary problem we call the step a *unilateral reduction.* There are two kinds of unilateral reduction, and both are, in some way or other, more risky than a bilateral or convertible reduction.

Our example 2 shows a unilateral reduction to a less ambitious problem. In fact, if we could solve the original problem, concerned with a parallelepiped whose length, width, and height are a, b, c respectively, we could move on to the auxiliary problem putting $c = 0$ and obtaining a parallelogram with length a and width b. For another example of a unilateral reduction to a less ambitious problem see SPECIALIZATION, 3, 4, 5. These examples show that, with some luck, we may be able to use a less ambitious auxiliary problem as a *stepping stone,* combining the solution of the auxiliary problem with some appropriate supplementary remark to obtain the solution of the original problem.

Unilateral reduction to a more ambitious problem may also be successful. (See GENERALIZATION, 2, and the reduction of the first to the second problem considered in INDUCTION AND MATHEMATICAL INDUCTION, 1, 2.) In fact, the more ambitious problem may be more accessible; this is the INVENTOR'S PARADOX.

Bolzano, Bernard (1781-1848), logician and mathematician, devoted an extensive part of his comprehensive presentation of logic, Wissenschaftslehre, to the subject of heuristic (vol. 3, pp. 293-575). He writes about this part of his work: "I do not think at all that I am able to present here any procedure of investigation that was not perceived long ago by all men of talent; and I do not promise at all that you can find here anything quite new of this kind. But I shall take pains to state in clear words the rules and ways of investigation which are followed by all able men, who in most cases are not even conscious of following them. Although I am free from the illusion that I shall fully succeed even in doing this, I still hope that the little that is presented here may please some people and have some application afterwards."

Bright idea, or "good idea," or "seeing the light," is a colloquial expression describing a sudden advance toward the solution; see PROGRESS AND ACHIEVEMENT, 6. The coming of a bright idea is an experience familiar to everybody but difficult to describe and so it may be interesting to notice that a very suggestive description of it has been incidentally given by an authority as old as Aristotle.

Most people will agree that conceiving a bright idea is an "act of sagacity." Aristotle defines "sagacity" as follows: "Sagacity is a hitting by guess upon the essential connection in an inappreciable time. As for example, if you see a person talking with a rich man in a certain way, you may instantly guess that that person is trying to borrow money. Or observing that the bright side of the moon is always toward the sun, you may suddenly perceive why this is; namely, because the moon shines by the light of the sun."[1]

The first example is not bad but rather trivial; not much sagacity is needed to guess things of this sort about rich men and money, and the idea is not very bright. The second example, however, is quite impressive if we make a little effort of imagination to see it in its proper setting.

We should realize that a contemporary of Aristotle had to watch the sun and the stars if he wished to know the time since there were no wristwatches, and had to observe the phases of the moon if he planned traveling by night since there were no street lights. He was much better acquainted with the sky than the modern city-

[1] The text is slightly rearranged. For a more exact translation see William Whewell, *The Philosophy of the Inductive Sciences* (1847), vol. II, p. 131.

dweller, and his natural intelligence was not dimmed by undigested fragments of journalistic presentations of astronomical theories. He saw the full moon as a flat disc, similar to the disc of the sun but much less bright. He must have wondered at the incessant changes in the shape and position of the moon. He observed the moon occasionally also at daytime, about sunrise or sunset, and found out "that the bright side of the moon is always toward the sun" which was in itself a respectable achievement. And now he perceives that the varying aspects of the moon are like the various aspects of a ball which is illuminated from one side so that one half of it is shiny and the other half dark. He conceives the sun and the moon not as flat discs but as round bodies, one giving and the other receiving the light. He understands the essential connection, he rearranges his former conceptions instantly, "in an inappreciable time": there is a sudden leap of the imagination, a bright idea, a flash of genius.

Can you check the result? *Can you check the argument?* A good answer to these questions strengthens our trust in the solution and contributes to the solidity of our knowledge.

1. Numerical results of mathematical problems can be tested by comparing them to observed numbers, or to a commonsense estimate of observable numbers. As problems arising from practical needs or natural curiosity almost always aim at facts it could be expected that such comparisons with observable facts are seldom omitted. Yet every teacher knows that students achieve incredible things in this respect. Some students are not disturbed at all when they find 16,130 ft. for the length of the boat and 8 years, 2 months for the age of the captain who is, by the way, known to be a grandfather. Such neglect of

the obvious does not show necessarily stupidity but rather indifference toward artificial problems.

2. Problems "in letters" are susceptible of more, and more interesting, tests than "problems in numbers" (section 14). For another example, let us consider the frustum of a pyramid with square base. If the side of the lower base is a, the side of the upper base b, and the altitude of the frustum h, we find for the volume

$$\frac{a^2 + ab + b^2}{3} h.$$

We may test this result by SPECIALIZATION. In fact, if $b = a$ the frustum becomes a prism and the formula yields a^2h; and if $b = 0$ the frustum becomes a pyramid and the formula yields $\frac{a^2h}{3}$. We may apply the TEST BY DIMENSION. In fact, the expression has as dimension the cube of a length. Again, we may test the formula by *variation of the data.* In fact, if any one of the positive quantities a, b or h increases the value of the expression increases.

Tests of this sort can be applied not only to the final result but also to intermediate results. They are so useful that it is worth while preparing for them; see VARIATION OF THE PROBLEM, 4. In order to be able to use such tests, we may find advantage in generalizing a "problem in numbers" and changing it into a "problem in letters"; see GENERALIZATION, 3.

3. *Can you check the argument?* Checking the argument step by step, we should avoid mere repetition. First, mere repetition is apt to become boring, uninstructive, a strain on the attention. Second, where we stumbled once, there we are likely to stumble again if the circumstances are the same as before. If we feel that it is necessary to go again through the whole argument step by step, we should

at least change the order of the steps, or their grouping, to introduce some variation.

4. It requires less exertion and is more interesting to pick out the weakest point of the argument and examine it first. A question very useful in picking out points of the argument that are worth while examining is: DID YOU USE ALL THE DATA?

5. It is clear that our nonmathematical knowledge cannot be based entirely on formal proofs. The more solid part of our everyday knowledge is continually tested and strengthened by our everyday experience. Tests by observation are more systematically conducted in the natural sciences. Such tests take the form of careful experiments and measurements, and are combined with mathematical reasoning in the physical sciences. Can our knowledge in mathematics be based on formal proofs alone?

This is a philosophical question which we cannot debate here. It is certain that your knowledge, or my knowledge, or your students' knowledge in mathematics is not based on formal proofs alone. If there is any solid knowledge at all, it has a broad experimental basis, and this basis is broadened by each problem whose result is successfully tested.

Can you derive the result differently? When the solution that we have finally obtained is long and involved, we naturally suspect that there is some clearer and less roundabout solution: *Can you derive the result differently? Can you see it at a glance?* Yet even if we have succeeded in finding a satisfactory solution we may still be interested in finding another solution. We desire to convince ourselves of the validity of a theoretical result by two different derivations as we desire to perceive a material object through two different senses. Having found a proof, we wish to find another proof as we wish to touch an object after having seen it.

Two proofs are better than one. "It is safe riding at two anchors."

1. *Example.* Find the area S of the lateral surface of the frustum of a right circular cone, being given the radius of the lower base R, the radius of the upper base r, and the altitude h.

This problem can be solved by various procedures. For instance, we may know the formula for the lateral surface of a full cone. As the frustum is generated by cutting off from a cone a smaller cone, so its lateral surface is the difference of two full conical surfaces; it remains to express these in terms of R, r, h. Carrying through this idea, we obtain finally the formula

$$S = \pi(R + r)\sqrt{(R - r)^2 + h^2}.$$

Having found this result in some way or other, after longer calculation, we may desire a clearer and less roundabout argument. *Can you derive the result differently? Can you see it at a glance?*

Desiring to see intuitively the whole result, we may begin with trying to see the geometric meaning of its parts. Thus, we may observe that

$$\sqrt{(R - r)^2 + h^2}$$

is the length of the *slant height.* (The slant height is one of the nonparallel sides of the isosceles trapezoid that, revolving about the line joining the midpoints of its parallel sides, generates the frustum; see Fig. 12.) Again, we may discover that

$$\pi(R + r) = \frac{2\pi R + 2\pi r}{2}$$

is the arithmetic mean of the perimeters of the two bases

of the frustum. Looking at the same part of the formula, we may be moved to write it also in the form

$$\pi(R + r) = 2\pi \frac{R + r}{2}$$

that is the *perimeter of the mid-section* of the frustum. (We call here mid-section the intersection of the frustum with a plane which is parallel both to the lower base and to the upper base of the frustum and bisects the altitude.)

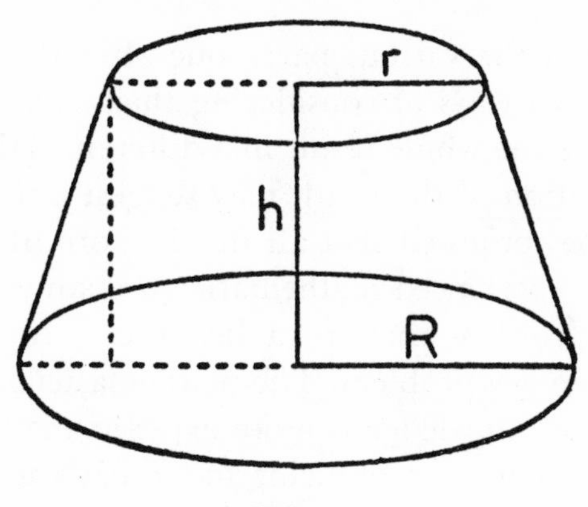

FIG. 12

Having found new interpretations of various parts, we may see now the whole formula in a different light. We may read it thus:

Area = Perimeter of mid-section × Slant height.

We may recall here the rule for the trapezoid:

Area = Middle-line × Altitude.

(The middle-line is parallel to the two parallel sides of the trapezoid and bisects the altitude.) Seeing intuitively the analogy of both statements, that about the frustum and that about the trapezoid, we see the whole result about the frustum "almost at a glance." That is, we feel that we are very near now to a short and direct proof of the result found by a long calculation.

2. The foregoing example is typical. Not entirely satisfied with our derivation of the result, we wish to improve it, to change it. Therefore, we study the result, trying to understand it better, to see some new aspect of it. We may succeed first in observing a new interpretation of a certain small part of the result. Then, we may be lucky enough to discover some new mode of conceiving some other part.

Examining the various parts, one after the other, and trying various ways of considering them, we may be led finally to see the whole result in a different light, and our new conception of the result may suggest a new proof.

It may be confessed that all this is more likely to happen to an experienced mathematician dealing with some advanced problem than to a beginner struggling with some elementary problem. The mathematician who has a great deal of knowledge is more exposed than the beginner to the danger of mobilizing too much knowledge and framing an unnecessarily involved argument. But, as a compensation, the experienced mathematician is in a better position than the beginner to appreciate the reinterpretation of a small part of the result and to proceed, accumulating such small advantages, to recasting ultimately the whole result.

Nevertheless, it can happen even in very elementary classes that the students present an unnecessarily complicated solution. Then, the teacher should show them, at least once or twice, not only how to solve the problem more shortly but also how to find, in the result itself, indications of a shorter solution.

See also REDUCTIO AD ABSURDUM AND INDIRECT PROOF.

Can you use the result? To find the solution of a problem by our own means is a discovery. If the problem is not difficult, the discovery is not so momentous, but it is a discovery nevertheless. Having made some discovery, however modest, we should not fail to inquire whether there is something more behind it, we should not miss the possibilities opened up by the new result, we should try to use again the procedure used. Exploit your success! *Can you use the result, or the method, for some other problem?*

1. We can easily imagine new problems if we are somewhat familiar with the principal means of varying a problem, as GENERALIZATION, SPECIALIZATION, ANALOGY, DECOMPOSING AND RECOMBINING. We start from a proposed problem, we derive from it others by the means we just mentioned, from the problems we obtained we derive still others, and so on. The process is unlimited in theory but, in practice, we seldom carry it very far, because the problems that we obtain so are apt to be inaccessible.

On the other hand we can construct new problems which we can easily solve using the solution of a problem previously solved; but these easy new problems are apt to be uninteresting.

To find a new problem which is both interesting and accessible, is not so easy; we need experience, taste, and good luck. Yet we should not fail to look around for more good problems when we have succeeded in solving one. Good problems and mushrooms of certain kinds have something in common; they grow in clusters. Having found one, you should look around; there is a good chance that there are some more quite near.

2. We are going to illustrate some of the foregoing points by the same example that we discussed in sections **8, 10, 12, 14, 15.** Thus we start from the following problem:

Given the three dimensions (length, breadth, and height) of a rectangular parallelepiped, find the diagonal.

If we know the solution of this problem, we can easily

solve any of the following problems (of which the first two were almost stated in section **14**).

Given the three dimensions of a rectangular parallelepiped, find the radius of the circumscribed sphere.

The base of a pyramid is a rectangle of which the center is the foot of the altitude of the pyramid. Given the altitude of the pyramid and the sides of its base, find the lateral edges.

Given the rectangular coordinates (x_1, y_1, z_1), (x_2, y_2, z_2) of two points in space, find the distance of these points.

We solve these problems easily because they are scarcely different from the original problem whose solution we know. In each case, we add some new notion to our original problem, as circumscribed sphere, pyramid, rectangular coordinates. These notions are easily added and easily eliminated, and, having got rid of them, we fall back upon our original problem.

The foregoing problems have a certain interest because the notions that we introduced into the original problem are interesting. The last problem, that about the distance of two points given by their coordinates, is even an important problem because rectangular coordinates are important.

3. Here is another problem which we can easily solve if we know the solution of our original problem: Given the length, the breadth, and the diagonal of a rectangular parallelepiped, find the height.

In fact, the solution of our original problem consists essentially in establishing a relation among four quantities, the three dimensions of the parallelepiped and its diagonal. If any three of these four quantities are given, we can calculate the fourth from the relation. Thus, we can solve the new problem.

We have here a pattern to derive easily solvable new

problems from a problem we have solved: we regard the original unknown as given and one of the original data as unknown. The relation connecting the unknown and the data is the same in both problems, the old and the new. Having found this relation in one, we can use it also in the other.

This pattern of deriving new problems by interchanging the roles is very different from the pattern followed under 2.

4. Let us now derive some new problems by other means.

A natural *generalization* of our original problem is the following: Find the diagonal of a parallelepiped, being given the three edges issued from an end-point of the diagonal, and the three angles between these three edges.

By *specialization* we obtain the following problem: Find the diagonal of a cube with given edge.

We may be led to an inexhaustible variety of problems by *analogy*. Here are a few derived from those considered under 2: Find the diagonal of a regular octahedron with given edge. Find the radius of the circumscribed sphere of a regular tetrahedron with given edge. Given the geographical coordinates, latitude and longitude, of two points on the earth's surface (which we regard as a sphere) find their spherical distance.

All these problems are interesting but only the one obtained by specialization can be solved immediately on the basis of the solution of the original problem.

5. We may derive new problems from a proposed one by considering certain of its elements as variable.

A special case of a problem mentioned under 2 is to find the radius of a sphere circumscribed about a cube whose edge is given. Let us regard the cube, and the common center of cube and sphere as fixed, but let us vary the radius of the sphere. If this radius is small, the sphere is contained in the cube. As the radius increases, the sphere expands (as a rubber balloon in the process of being inflated). At a certain moment, the sphere touches the faces of the cube; a little later, its edges; still later the sphere passes through the vertices. Which values does the radius assume at these three critical moments?

6. The mathematical experience of the student is incomplete if he never had an opportunity to solve a *problem invented by himself.* The teacher may show the derivation of new problems from one just solved and, doing so, provoke the curiosity of the students. The teacher may also leave some part of the invention to the students. For instance, he may tell about the expanding sphere we just discussed (under 5) and ask: "What would you try to calculate? Which value of the radius is particularly interesting?"

Carrying out. To conceive a plan and to carry it through are two different things. This is true also of mathematical problems in a certain sense; between carrying out the plan of the solution, and conceiving it, there are certain differences in the character of the work.

1. We may use provisional and merely plausible arguments when devising the final and rigorous argument as we use scaffolding to support a bridge during construction. When, however, the work is sufficiently advanced we take off the scaffolding, and the bridge should be able to stand by itself. In the same way, when the solution is sufficiently advanced, we brush aside all kinds of provisional and merely plausible arguments, and the result should be supported by rigorous argument alone.

Devising the plan of the solution, we should not be too afraid of merely plausible, heuristic reasoning. Anything is right that leads to the right idea. But we have to change this standpoint when we start carrying out the plan and then we should accept only conclusive, strict arguments. *Carrying out your plan of the solution check each step. Can you see clearly that the step is correct?*

The more painstakingly we check our steps when carrying out the plan, the more freely we may use heuristic reasoning when devising it.

2. We should give some consideration to the order in which we work out the details of our plan, especially if our problem is complex. We should not omit any detail, we should understand the relation of the detail before us to the whole problem, we should not lose sight of the connection of the major steps. Therefore, we should proceed in proper order.

In particular, it is not reasonable to check minor details before we have good reasons to believe that the major steps of the argument are sound. If there is a break in the main line of the argument, checking this or that secondary detail would be useless anyhow.

The order in which we work out the details of the argument may be very different from the order in which we invented them; and the order in which we write down the details in a definitive exposition may be still different. Euclid's Elements present the details of the argument in a rigid systematic order which was often imitated and often criticized.

3. In Euclid's exposition all arguments proceed in the same direction: from the data toward the unknown in "problems to find," and from the hypothesis toward the conclusion in "problems to prove." Any new element, point, line, etc., has to be correctly derived from the data or from elements correctly derived in foregoing steps. Any new assertion has to be correctly proved from the hypothesis or from assertions correctly proved in foregoing steps. Each new element, each new assertion is

examined when it is encountered first, and so it has to be examined just once; we may concentrate all our attention upon the present step, we need not look behind us, or look ahead. The very last new element whose derivation we have to check, is the unknown. The very last assertion whose proof we have to examine, is the conclusion. If each step is correct, also the last one, the whole argument is correct.

The Euclidean way of exposition can be highly recommended, without reservation, if the purpose is to examine the argument in detail. Especially, if it is our own argument, and it is long and complicated, and we have not only found it but have also surveyed it on large lines so that nothing is left but to examine each particular point in itself, then nothing is better than to write out the whole argument in the Euclidean way.

The Euclidean way of exposition, however, cannot be recommended without reservation if the purpose is to convey an argument to a reader or to a listener who never heard of it before. The Euclidean exposition is excellent to show each particular point but not so good to show the main line of the argument. THE INTELLIGENT READER can easily see that each step is correct but has great difficulty in perceiving the source, the purpose, the connection of the whole argument. The reason for this difficulty is that the Euclidean exposition fairly often proceeds in an order exactly opposite to the natural order of invention. (Euclid's exposition follows rigidly the order of "synthesis"; see PAPPUS, especially comments 3, 4, 5.)

4. Let us sum up. Euclid's manner of exposition, progressing relentlessly from the data to the unknown and from the hypothesis to the conclusion, is perfect for checking the argument in detail but far from being perfect for making understandable the main line of the argument.

It is highly desirable that the students should examine their own arguments in the Euclidean manner, proceeding from the data to the unknown, and checking each step although nothing of this kind should be too rigidly enforced. It is not so desirable that the teacher should present many proofs in the pure Euclidean manner, although the Euclidean presentation may be very useful after a discussion in which, as is recommended by the present book, the students guided by the teacher discover the main idea of the solution as independently as possible. Also desirable seems to be the manner adopted by some textbooks in which an intuitive sketch of the main idea is presented first and the details in the Euclidean order of exposition afterwards.

5. Wishing to satisfy himself that his proposition is true, the conscientious mathematician tries to see it intuitively and to give a formal proof. *Can you see clearly that it is correct? Can you prove that it is correct?* The conscientious mathematician acts in this respect like the lady who is a conscientious shopper. Wishing to satisfy herself of the quality of a fabric, she wants to see it and to touch it. Intuitive insight and formal proof are two different ways of perceiving the truth, comparable to the perception of a material object through two different senses, sight and touch.

Intuitive insight may rush far ahead of formal proof. Any intelligent student, without any systematic knowledge of solid geometry, can see as soon as he has clearly understood the terms that two straight lines parallel to the same straight line are parallel to each other (the three lines may or may not be in the same plane). Yet the proof of this statement, as given in proposition 9 of the 11th book of Euclid's Elements, needs a long, careful, and ingenious preparation.

Formal manipulation of logical rules and algebraic formulas may get far ahead of intuition. Almost everybody can see at once that 3 straight lines, taken at random, divide the plane into 7 parts (look at the only finite part, the triangle included by the 3 lines). Scarcely anybody is able to see, even straining his attention to the utmost, that 5 planes, taken at random, divide space into 26 parts. Yet it can be rigidly proved that the right number is actually 26, and the proof is not even long or difficult.

Carrying out our plan, we check each step. Checking our step, we may rely on intuitive insight or on formal rules. Sometimes the intuition is ahead, sometimes the formal reasoning. It is an interesting and useful exercise to do it both ways. *Can you see clearly that the step is correct?* Yes, I can see it clearly and distinctly. Intuition is ahead; but could formal reasoning overtake it? *Can you also* PROVE *that it is correct?*

Trying to prove formally what is seen intuitively and to see intuitively what is proved formally is an invigorating mental exercise. Unfortunately, in the classroom there is not always enough time for it. The example, discussed in sections **12** and **14**, is typical in this respect.

Condition is a principal part of a "problem to find." See PROBLEMS TO FIND, PROBLEMS TO PROVE, 3. See also TERMS, NEW AND OLD, 2.

A condition is called *redundant* if it contains superfluous parts. It is called *contradictory* if its parts are mutually opposed and inconsistent so that there is no object satisfying the condition.

Thus, if a condition is expressed by more linear equations than there are unknowns, it is either redundant or contradictory; if the condition is expressed by fewer equations than there are unknowns, it is insufficient to determine the unknowns; if the condition is expressed by just as many equations as there are unknowns it is usually just sufficient to determine the unknowns but may be, in exceptional cases, contradictory or insufficient.

Contradictory. See CONDITION.

Corollary is a theorem which we find easily in examining another theorem just found. The word is of Latin origin; a more literal translation would be "gratuity" or "tip."

Could you derive something useful from the data? We have before us an unsolved problem, an open question. We have to *find the connection between the data and the unknown.* We may represent our unsolved problem as open space between the data and the unknown, as a gap across which we have to construct a bridge. We can start constructing our bridge from either side, from the unknown or from the data.

Look at the unknown! And try to think of a familiar problem having the same or a similar unknown. This suggests starting the work from the unknown.

Look at the data! *Could you derive something useful from the data?* This suggests starting the work from the data.

It appears that starting the reasoning from the unknown is usually preferable (see PAPPUS and WORKING BACKWARDS). Yet the alternative start, from the data, also has chances of success, must often be tried, and deserves illustration.

Example. We are given three points *A, B,* and *C*. Draw a line through *A* which passes between *B* and *C* and is at equal distances from *B* and *C*.

What are the data? Three points, *A, B,* and *C,* are given in position. We *draw a figure,* exhibiting the data (Fig. 13).

What is the unknown? A straight line.

What is the condition? The required line passes through A, and passes between B and C, at the same distance from each. We assemble the unknown and the data

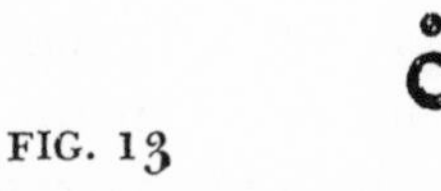

FIG. 13

in a figure exhibiting the required relations (Fig. 14). Our figure, suggested by the *definition* of the distance of a point from a straight line, shows the right angles involved by this definition.

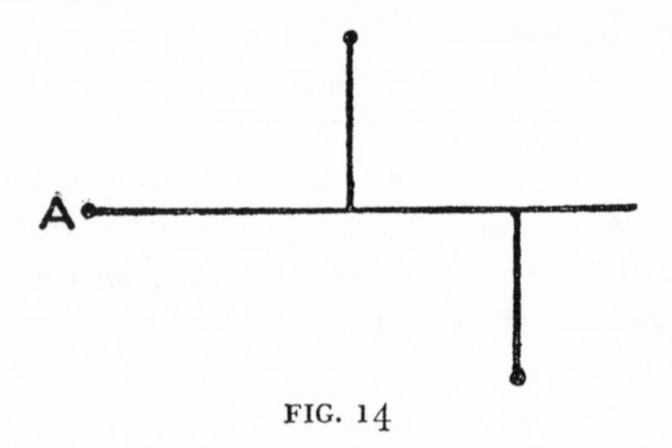

FIG. 14

The figure, as it is plotted, is still "too empty." The unknown straight line is still unsatisfactorily connected with the data A, B, and C. The figure needs some auxiliary line, some addition—but what? A fairly good student can get stranded here. There are, of course, various things to try, but the best question to refloat him is: *Could you derive something useful from the data?*

In fact, what are the data? The three points exhibited in Fig. 13, nothing else. We have not yet used sufficiently the points B and C; we have to derive something useful from them. But what can you do with just two points? Join them by a straight line. So, we draw Fig. 15.

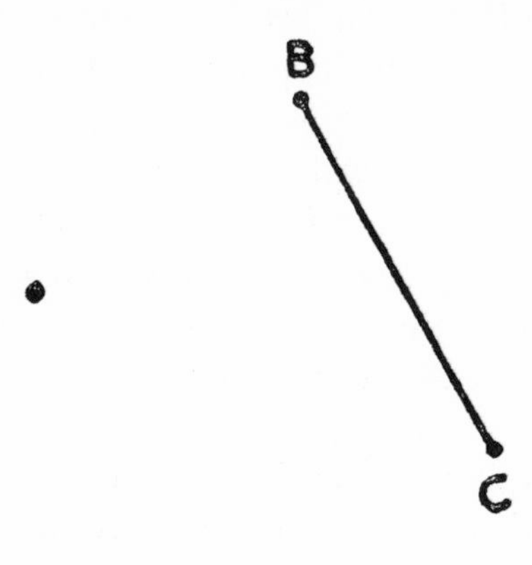

FIG. 15

If we superpose Fig. 14 and Fig. 15, the solution may appear in a flash: There are two right triangles, they are congruent, there is an all-important new point of intersection.

Could you restate the problem? *Could you restate it still differently?* These questions aim at suitable VARIATION OF THE PROBLEM.

Go back to definitions. See DEFINITION.

Decomposing and recombining are important operations of the mind.

You examine an object that touches your interest or challenges your curiosity: a house you intend to rent, an important but cryptic telegram, any object whose purpose and origin puzzle you, or any problem you intend to solve. You have an impression of the object as a whole but this impression, possibly, is not definite enough. A detail strikes you, and you focus your attention upon it. Then, you concentrate upon another detail; then, again, upon another. Various combinations of details may present themselves and after a while you again consider the object as a whole but you see it now differently. You decompose the whole into its parts, and you recombine the parts into a more or less different whole.

1. If you go into detail you may lose yourself in details. Too many or too minute particulars are a burden on the mind. They may prevent you from giving sufficient attention to the main point, or even from seeing the main point at all. Think of the man who cannot see the forest for the trees.

Of course, we do not wish to waste our time with unnecessary detail and we should reserve our effort for the essential. The difficulty is that we cannot say beforehand which details will turn out ultimately as necessary and which will not.

Therefore, let us, first of all, understand the problem as a whole. Having understood the problem, we shall be in a better position to judge which particular points may be the most essential. Having examined one or two essential points we shall be in a better position to judge which further details might deserve closer examination. Let us go into detail and decompose the problem gradually, but not further than we need to.

Of course, the teacher cannot expect that all students should act wisely in this respect. On the contrary, it is a very foolish and bad habit with some students to start working at details before having understood the problem as a whole.

2. We are going to consider mathematical problems, "problems to find."

Having understood the problem as a whole, its aim, its main point, we wish to go into detail. Where should we start? In almost all cases, it is reasonable to begin with the consideration of the principal parts of the problem which are the unknown, the data, and the condition. In almost all cases it is advisable to start the detailed examination of the problem with the questions: *What is the unknown? What are the data? What is the condition?*

If we wish to examine further details, what should we do? Fairly often, it is advisable to examine each datum by itself, to *separate the various parts of the condition,* and to examine each part by itself.

We may find it necessary, especially if our problem is more difficult, to decompose the problem still further, and to examine still more remote details. Thus, it may be necessary to *go back to the definition* of a certain term, to introduce new elements involved by the definition, and to examine the elements so introduced.

3. After having decomposed the problem, we try to recombine its elements in some new manner. Especially, we may try to recombine the elements of the problem into some new, more accessible problem which we could possibly use as an auxiliary problem.

There are, of course, unlimited possibilities of recombination. Difficult problems demand hidden, exceptional, original combinations, and the ingenuity of the problem-solver shows itself in the originality of the combination. There are, however, certain usual and relatively simple sorts of combinations, sufficient for simpler problems, which we should know thoroughly and try first, even if we may be obliged eventually to resort to less obvious means.

There is a formal classification in which the most usual

and useful combinations are neatly placed. In constructing a new problem from the proposed problem, we may

(1) keep the unknown and change the rest (the data and the condition); or

(2) keep the data and change the rest (the unknown and the condition); or

(3) change both the unknown and the data.

We are going to examine these cases.

[The cases (1) and (2) overlap. In fact, it is possible to keep both the unknown and the data, and transform the problem by changing the form of the condition alone. For instance, the two following problems, although visibly equivalent, are not exactly the same:

Construct an equilateral triangle, being given a side.

Construct an equiangular triangle, being given a side.

The difference of the two statements which is slight in the present example may be momentous in other cases. Such cases are even important in certain respects but it would take up too much space to discuss them here. Compare AUXILIARY PROBLEMS, 7, last remark.]

4. *Keeping the unknown* and changing the data and the condition in order to transform the proposed problem is often useful. The suggestion LOOK AT THE UNKNOWN aims at problems with the same unknown. We may try to recollect a formerly solved problem of this kind: *And try to think of a familiar problem having the same or a similar unknown.* Failing to remember such a problem we may try to invent one: *Could you think of other data appropriate to determine the unknown?*

A new problem which is more closely related to the proposed problem has a better chance of being useful. Therefore, keeping the unknown, we try to keep also some data and some part of the condition, and to change, as little as feasible, only one or two data and a small part of the condition. A good method is one in which we omit something without adding anything; we keep the unknown, *keep only a part of the condition, drop the other part,* but do not introduce any new clause or datum. Examples and comments on this case follow under 7, 8.

5. *Keeping the data,* we may try to introduce some useful and more accessible new unknown. Such an unknown must be obtained from the original data and we have such an unknown in mind when we ask: COULD YOU DERIVE SOMETHING USEFUL FROM THE DATA?

Let us observe that two things are here desirable. First, the new unknown should be more accessible, that is, more easily obtainable from the data than the original unknown. Second, the new unknown should be useful, that is, it should be, when found, capable of rendering some definite service in the search of the original unknown. In short, the new unknown should be a sort of *stepping stone.* A stone in the middle of the creek is nearer to me than the other bank which I wish to arrive at and, when the stone is reached, it helps me on toward the other bank.

The new unknown should be both accessible and useful but, in practice, we must often content ourselves with less. If nothing better presents itself, it is not unreasonable to derive something from the data that has some chance of being useful; and it is also reasonable to try a new unknown which is closely connected with the original one, even if it does not seem particularly accessible from the outset.

For instance, if our problem is to find the diagonal of a parallelepiped (as in section 8) we may introduce the diagonal of a face as new unknown. We may do so either because we *know* that if we have the diagonal of the face we can also obtain the diagonal of the solid (as in section 10); or we may do so because we see that the diagonal of the face is easy to obtain and we *suspect* that it might be useful in finding the diagonal of the solid. (Compare DID YOU USE ALL THE DATA? 1.)

If our problem is to construct a circle, we have to find two things, its center and its radius; our problem has two parts, we may say. In certain cases, one part is more accessible than the other and therefore, in any case, we may reasonably give a moment's consideration to this possibility: *Could you solve a part of the problem?* Asking this, we weigh the chances: Would it pay to concentrate just upon the center, or just upon the radius, and to choose one or the other as our new unknown? Questions of this sort are very often useful. In more complex or in more advanced problems, the decisive idea often consists in carving out some more accessible but essential part from the problem.

6. *Changing both the unknown and the data* we deviate more from our original course than in the foregoing cases. This, naturally, we do not like; we sense the danger of losing the original problem altogether. Yet we may be compelled to such an extensive change if less radical changes have failed to produce something accessible and useful, and we may be tempted to recede so far from our original problem if the new problem has a good chance of success. *Could you change the unknown, or the data, or both if necessary, so that the new unknown and the new data are nearer to each other?*

An interesting way of changing both the unknown and the data is interchanging the unknown with one of the data. (See CAN YOU USE THE RESULT? 3.)

7. *Example.* Construct a triangle, being given a side a, the altitude h perpendicular to a, and the angle α opposite to a.

What is the unknown? A triangle.

What are the data? Two lines, a and h, and an angle α.

Now, if we are somewhat familiar with problems of

geometric construction, we try to reduce such a problem to the construction of a point. We draw a line BC equal to the given side a; then all that we have to find is the vertex of the triangle A, opposite to a, see Fig. 16. We have, in fact, a new problem.

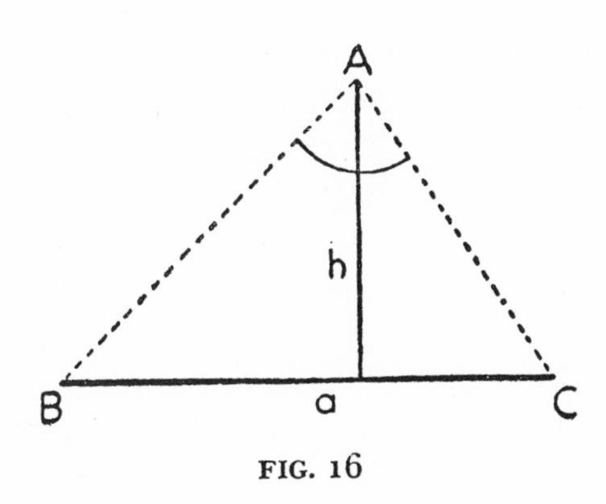

FIG. 16

What is the unknown? The point A.

What are the data? A line h, an angle α, and two points B and C given in position.

What is the condition? The perpendicular distance of the point A from the line BC should be h and $\angle BAC = \alpha$.

In fact, we have transformed our problem, changing both the unknown and the data. The new unknown is a point, the old unknown was a triangle. Some of the data are the same in both problems, the line h and the angle α; but in the old problem we were given a length a and now we are given two points, B and C, instead.

The new problem is not difficult. The following suggestion brings us quite near to the solution.

Separate the various parts of the condition. The condition has two parts, one concerned with the datum h, the other with the datum α. The unknown point is required to be

(I) at distance h from the line BC; and

(II) the vertex of an angle of given magnitude α, whose sides pass through the given points B and C.

If we *keep only one part of the condition and drop the other part,* the unknown point is not completely determined. There are many points satisfying part (I) of the condition, namely all points of a parallel to the line BC at the distance h from BC.[2] This parallel is the locus of the points satisfying part (I) of the condition. The locus of the points satisfying part (II) is a certain circular arc whose end-points are B and C. We can describe both loci; their intersection is the point that we desired to construct.

The procedure that we have just applied has a certain interest; solving problems of geometric construction, we can often follow successfully its pattern: Reduce the problem to the construction of a point, and construct the point as an intersection of two loci.

But a certain step of this procedure has a still more general interest; solving "problems to find" of any kind, we can follow its pattern: *Keep only a part of the condition, drop the other part.* Doing so, we weaken the condition of the proposed problem, we restrict less the unknown. *How far is the unknown then determined, how can it vary?* By asking this, we set, in fact, a new problem. If the unknown is a point in the plane (as it was in our example) the solution of this new problem consists in determining a locus described by the point. If the unknown is a mathematical object of some other kind (it was a square in section 18) we have to describe properly and to characterize precisely a certain set of objects. Even if the unknown is not a mathematical

[2] The plane is bisected by the line through B and C. We choose one of the halfplanes to construct A in it, and so we may consider just one parallel to BC; otherwise, we should consider two such parallels.

object (as in the next example, under 8) it may be useful to consider, to characterize, to describe, or to list those objects which satisfy a certain part of the condition imposed upon the unknown by the proposed problem.

8. *Example.* In a crossword puzzle that allows puns and anagrams we find the following clue:

"Forward and backward part of a machine (5 letters)."

What is the unknown? A word.

What is the condition? The word has 5 letters. It has something to do with some part of some machine. It should be, of course, an English word, and not a too unusual one, let us hope.

Is the condition sufficient to determine the unknown? No. Or, rather, the condition may be sufficient but that part of the condition which is clear by now is certainly insufficient. There are too many words satisfying it, as "lever," or "screw," or what not.

The condition is ambiguously expressed—on purpose, of course. If nothing can be found that could be plausibly described as a "forward part" of a machine and would be a "backward part" too, we may suspect that forward and backward *reading* might be meant. It may be a good idea to examine this interpretation of the clue.

Separate the various parts of the condition. The condition has two parts, one concerned with the meaning of the word, the other with its spelling. The unknown word is required to be

(I) a short word meaning some part of some machine;

(II) a word with 5 letters which spelled backward give again a word meaning some part of some machine.

If we *keep only one part of the condition and drop the other part,* the unknown is not completely determined. There are many words satisfying part (I) of the condition, we have a sort of locus. We may "describe" this locus (I), "follow" it to its "intersection" with locus

(II). The natural procedure is to concentrate upon part (I) of the condition, to recollect words having the prescribed meaning and, when we have succeeded in recollecting some such word, to examine whether it has or has not the prescribed length and can or cannot be read backward. We may have to recollect several words before we run into the right one: lever, screw, wheel, shaft, hinge, motor.

Of course, "rotor"!

9. Under 3, we classified the possibilities of obtaining a new "problem to find" by recombining certain elements of a proposed "problem to find." If we do not introduce just one new problem, but two or more new problems, there are more possibilities which we have to mention but do not attempt to classify.

Still other possibilities may arise. Especially, the solution of a "problem to find" may depend on the solution of a "problem to prove." We just mention this important possibility; considerations of space prevent us from discussing it.

10. Only few and short remarks can be added concerning "problems to prove"; they are analogous to the foregoing more extensive comments on "problems to find" (2 to 9).

Having understood such a problem as a whole, we should, in general, examine its principal parts. The principal parts are the hypothesis and the conclusion of the theorem that we are required to prove or to disprove. We should understand these parts thoroughly: *What is the hypothesis? What is the conclusion?* If there is need to get down to more particular points, we may *separate the various parts of the hypothesis,* and consider each part by itself. Then we may proceed to other details, decomposing the problem further and further.

After having decomposed the problem, we may try to recombine its elements in some new manner. Especially, we may try to recombine the elements into another theorem. In this respect, there are three possibilities.

(1) We *keep the conclusion* and change the hypothesis. We first try to recollect such a theorem: *Look at the conclusion! And try to think of a familiar theorem having the same or a similar conclusion.* If we do not succeed in recollecting such a theorem we try to invent one: *Could you think of another hypothesis from which you could easily derive the conclusion?* We may change the hypothesis by omitting something without adding anything: *Keep only a part of the hypothesis, drop the other part; is the conclusion still valid?*

(2) We *keep the hypothesis* and change the conclusion: *Could you derive something useful from the hypothesis?*

(3) We *change both the hypothesis and the conclusion.* We may be more inclined to change both if we have had no success in changing just one. *Could you change the hypothesis, or the conclusion, or both if necessary, so that the new hypothesis and the new conclusion are nearer to each other?*

We do not attempt to classify here the various possibilities which arise when, in order to solve the proposed "problem to prove," we introduce two or more new "problems to prove," or when we link it up with an appropriate "problem to find."

Definition of a term is a statement of its meaning in other terms which are supposed to be well known.

1. *Technical terms* in mathematics are of two kinds. Some are accepted as primitive terms and are not defined. Others are considered as derived terms and are defined in due form; that is, their meaning is stated in primitive

terms and in formerly defined derived terms. Thus, we do not give a formal definition of such primitive notions as point, straight line, and plane.[3] Yet we give formal definitions of such notions as "bisector of an angle" or "circle" or "parabola."

The definition of the last quoted term may be stated as follows. We call *parabola* the locus of points which are at equal distance from a fixed point and a fixed straight line. The fixed point is called the *focus* of the parabola, the fixed line its *directrix*. It is understood that all elements considered are in a fixed plane, and that the fixed point (the focus) is not on the fixed line (the directrix).

The reader is not supposed to know the meaning of the terms defined: parabola, focus of the parabola, directrix of the parabola. But he is supposed to know the meaning of all the other terms as point, straight line, plane, distance of a point from another point, fixed, locus, etc.

2. *Definitions in dictionaries* are not very much different from mathematical definitions in the outward form but they are written in a different spirit.

The writer of a dictionary is concerned with the current meaning of the words. He *accepts,* of course, the current meaning and states it as neatly as he can in form of a definition.

The mathematician is not concerned with the current meaning of his technical terms, at least not primarily concerned with that. What "circle" or "parabola" or other technical terms of this kind may or may not denote in ordinary speech matters little to him. The mathematical definition *creates* the mathematical meaning.

[3] In this respect, ideas have changed since the time of Euclid and his Greek followers who defined the point, the straight line, and the plane. Their "definitions" however are scarcely formal definitions, rather intuitive illustrations of a sort. Illustrations, of course, are allowed, and even very desirable in teaching.

3. *Example.* Construct the point of intersection of a given straight line and a parabola of which the focus and the directrix are given.

Our approach to any problem must depend on the state of our knowledge. Our approach to the present problem depends mainly on the extent of our acquaintance with the properties of the parabola. If we know much about the parabola we try to make use of our knowledge and to extract something helpful from it: *Do you know a theorem that could be useful? Do you know a related problem?* If we know little about parabola, focus, and directrix, these terms are rather embarrassing and we naturally wish to get rid of them. How can we get rid of them? Let us listen to the dialogue of the teacher and the student discussing the proposed problem. They have chosen already a *suitable notation:* P for any of the unknown points of intersection, F for the focus, d for the directrix, c for the straight line intersecting the parabola.

"And *what is the unknown?*"

"The point P."

"*What are the data?*"

"The straight lines c and d, and the point F."

"*What is the condition?*"

"P is a point of intersection of the straight line c and of the parabola whose directrix is d and focus F."

"Correct. You had little opportunity, I know, to study the parabola but you can say, I think, what a parabola is."

"The parabola is the locus of points equidistant from the focus and the directrix."

"Correct. You remember the definition correctly. That is right, but we must also use it; *go back to definitions.* By virtue of the definition of the parabola, what can you say about your point P?"

"*P* is on the parabola. Therefore, *P* is equidistant from *d* and *F*."

"Good! *Draw a figure.*"

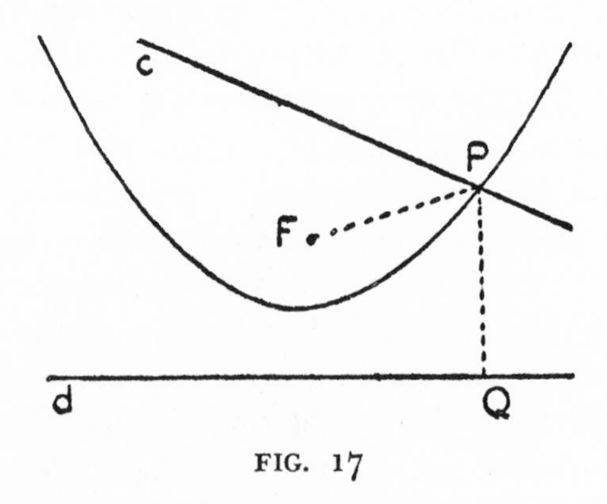

FIG. 17

The student introduces into Fig. 17 the lines *PF* and *PQ*, this latter being the perpendicular to *d* from *P*.

"Now, *could you restate the problem?*"

. . . . .

"Could you restate the condition of the problem, using the lines you have just introduced?"

"*P* is a point on the line *c* such that $PF = PQ$."

"Good. But please, say it in words: What is *PQ*?"

"The perpendicular distance of *P* from *d*."

"Good. Could you restate the problem now? But please, state it neatly, in a round sentence."

"Construct a point *P* on the given straight line *c* at equal distances from the given point *F* and the given straight line *d*."

"Observe the progress from the original statement to your restatement. The original statement of the problem was full of unfamiliar technical terms, parabola, focus, directrix; it sounded just a little pompous and inflated. And now, nothing remains of those unfamiliar technical terms; you have *deflated* the problem. Well done!"

4. *Elimination of technical terms* is the result of the work in the foregoing example. We started from a statement of the problem containing certain technical terms (parabola, focus, directrix) and we arrived finally at a restatement free of those terms.

In order to eliminate a technical term we must know its definition; but it is not enough to know the definition, we must use it. In the foregoing example, it was not enough to remember the definition of the parabola. The decisive step was to introduce into the figure the lines *PF* and *PQ* whose equality was granted by the definition of the parabola. This is the typical procedure. We introduce suitable elements into the conception of the problem. On the basis of the definition, we establish relations between the elements we introduced. If these relations express completely the meaning, we have used the definition. Having used its definition, we have eliminated the technical term.

The procedure just described may be called *going back to definitions.*

By going back to the definition of a technical term, we get rid of the term but introduce new elements and new relations instead. The resulting change in our conception of the problem may be important. At any rate, some restatement, some VARIATION OF THE PROBLEM is bound to result.

5. *Definitions and known theorems.* If we know the name "parabola" and have some vague idea of the shape of the curve but do not know anything else about it, our knowledge is obviously insufficient to solve the problem proposed as example, or any other serious geometric problem about the parabola. What kind of knowledge is needed for such a purpose?

The science of geometry may be considered as consisting of axioms, definitions, and theorems. The parabola is not mentioned in the axioms which deal only with such primitive terms as point, straight line, and so on. Any geometric argumentation concerned with the parabola, the solution of any problem involving it, must use either its definition or theorems about it. To solve such a problem, we must know, at least, the definition but it is better to know some theorems too.

What we said about the parabola is true, of course, of any derived notion. As we start solving a problem that involves such a notion, we cannot know yet what will be preferable to use, the definition of the notion, or some theorem about it; but it is certain that we have to use one or the other.

There are cases, however, in which we have no choice. If we know just the definition of the notion, and nothing else, then we are obliged to use the definition. If we do not know much more than the definition, our best chance may be to go back to the definition. But if we know many theorems about the notion, and have much experience in its use, there is some chance that we may get hold of a suitable theorem involving it.

6. *Several definitions.* The sphere is usually defined as the locus of points at a given distance from a given point. (The points are now in space, not restricted to a plane.) Yet the sphere could also be defined as the surface described by a circle revolving about a diameter. Still other definitions of the sphere are known, and many others possible.

When we have to solve a proposed problem involving some derived notion, as "sphere" or "parabola," and we wish to go back to its definition, we may have a choice among various definitions. Much may depend in such a case on choosing the definition that fits the case.

To find the area of the surface of the sphere was, at the time Archimedes solved it, a great and difficult problem.

Archimedes had the choice between the definitions of the sphere we just quoted. He preferred to conceive the sphere as the surface generated by a circle revolving about a fixed diameter. He inscribes in the circle a regular polygon, with an even number of sides, of which the fixed diameter joins opposite vertices. The regular polygon approximates the circle and, revolving with the circle, generates a convex surface composed of two cones with vertices at the extremities of the fixed diameter and of several frustums of cones in between. This composite surface approximates the sphere and is used by Archimedes in computing the area of the surface of the sphere. If we conceive the sphere as the locus of points equally distant from the center, no such simple approximation to its surface is suggested.

7. Going back to definitions is important in inventing an argument but it is also important in checking it.

Somebody presents an alleged new solution of Archimedes' problem of finding the area of the surface of the sphere. If he has only a vague idea of the sphere, his solution will not be any good. He may have a clear idea of the sphere but if he fails to use this idea in his argument I cannot know that he had any idea at all, and his argument is no good. Therefore, listening to the argument, I am waiting for the moment when he is going to say something substantial about the sphere, to use its definition or some theorem about it. If such a moment never comes, the solution is no good.

We should check not only the arguments of others but, of course, also our own arguments, in the same way. *Have you taken into account all essential notions involved in the problem?* How did you use this notion? Did you use its meaning, its definition? Did you use essential facts, known theorems about it?

That going back to definitions is important in examining the validity of an argument was emphasized by Pascal who stated the rule: "Substituer mentalement les définitions à la place des définis." The meaning is: "Substitute mentally the defining facts for the defined terms." That going back to definitions is also important in devising an argument was emphasized by Hadamard.

8. Going back to definitions is an important operation of the mind. If we wish to understand why the definitions of words are so important, we should realize first that words are important. We can hardly use our mind without using words, or signs, or symbols of some sort. Thus, words and signs have power. Primitive peoples believe that words and symbols have magic power. We may understand such belief but we should not share it. We should know that the power of a word does not reside in its sound, in the "vocis flatus," in the "hot air" produced by the speaker, but in the ideas of which the word reminds us and, ultimately, in the facts on which the ideas are based.

Therefore, it is a sound tendency to seek meaning and facts behind the words. Going back to definitions, the mathematician seeks to get hold of the actual relations of mathematical objects behind the technical terms, as the physicist seeks definite experiments behind his technical terms, and the common man with some common sense wants to get down to hard facts and not to be fooled by mere words.

Descartes, René (1596-1650), great mathematician and philosopher, planned to give a universal method to solve problems but he left unfinished his *Rules for the Direction of the Mind.* The fragments of this treatise, found in his manuscripts and printed after his death, contain more—and more interesting—materials concerning the solution of problems than his better known work *Dis-*

cours de la Méthode although the "Discours" was very likely written after the "Rules." The following lines of Descartes seem to describe the origin of the "Rules": "As a young man, when I heard about ingenious inventions, I tried to invent them by myself, even without reading the author. In doing so, I perceived, by degrees, that I was making use of certain rules."

Determination, hope, success. It would be a mistake to think that solving problems is a purely "intellectual affair"; determination and emotions play an important role. Lukewarm determination and sleepy consent to do a little something may be enough for a routine problem in the classroom. But, to solve a serious scientific problem, will power is needed that can outlast years of toil and bitter disappointments.

1. Determination fluctuates with hope and hopelessness, with satisfaction and disappointment. It is easy to keep on going when we think that the solution is just around the corner; but it is hard to persevere when we do not see any way out of the difficulty. We are elated when our forecast comes true. We are depressed when the way we have followed with some confidence is suddenly blocked, and our determination wavers.

"Il n'est point besoin espérer pour entreprendre ni réussir pour persévérer." "You can undertake without hope and persevere without success." Thus may speak an inflexible will, or honor and duty, or a nobleman with a noble cause. This sort of determination, however, would not do for the scientist, who should have some hope to start with, and some success to go on. In scientific work, it is necessary to apportion wisely determination to outlook. You do not take up a problem, unless it has some interest; you settle down to work seriously if the problem seems instructive; you throw in your whole personality if there is a great promise. If your purpose is set, you stick to it, but you do not make it unnecessarily difficult for yourself. You do not despise little successes, on the contrary, you seek them: *If you cannot solve the proposed problem try to solve first some related problem.*

2. When a student makes really silly blunders or is exasperatingly slow, the trouble is almost always the same; he has no desire at all to solve the problem, even no desire to understand it properly, and so he has not understood it. Therefore, a teacher wishing seriously to help the student should, first of all, stir up his curiosity, give him some desire to solve the problem. The teacher should also allow some time to the student to make up his mind, to settle down to his task.

Teaching to solve problems is education of the will. Solving problems which are not too easy for him, the student learns to persevere through unsuccess, to appreciate small advances, to wait for the essential idea, to concentrate with all his might when it appears. If the student had no opportunity in school to familiarize himself with the varying emotions of the struggle for the solution his mathematical education failed in the most vital point.

Diagnosis is used here as a technical term in education meaning "closer characterization of the student's work." A grade characterizes the student's work but somewhat crudely. The teacher, wishing to improve the student's work, needs a closer characterization of good and bad points as the physician, wishing to improve the patient's health, needs a diagnosis.

We are here particularly concerned with the student's efficiency in solving problems. How can we characterize it? We may derive some profit from the distinction of the four phases of the solution. In fact, the behavior of the students in the various phases is quite characteristic.

Incomplete *understanding of the problem,* owing to lack of concentration, is perhaps the most widespread deficiency in solving problems. With respect to *devising a plan* and obtaining a general idea of the solution two opposite faults are frequent. Some students rush into calculations and constructions without any plan or general idea; others wait clumsily for some idea to come and cannot do anything that would accelerate its coming. In *carrying out the plan,* the most frequent fault is carelessness, lack of patience in checking each step. Failure to *check the result* at all is very frequent; the student is glad to get an answer, throws down his pencil, and is not shocked by the most unlikely results.

The teacher, having made a careful diagnosis of a fault of this kind, has some chance to cure it by insisting on certain questions of the list.

Did you use all the data? Owing to the progressive mobilization of our knowledge, there will be much more in our conception of the problem at the end than was in it at the outset (PROGRESS AND ACHIEVEMENT, 1). But how is it now? Have we got what we need? Is our conception adequate? *Did you use all the data? Did you use the whole condition?* The corresponding question concerning "problems to prove" is: *Did you use the whole hypothesis?*

1. For an illustration, let us go back to the "parallelepiped problem" stated in section **8** (and followed up in sections **10, 12, 14, 15**). It may happen that a student runs into the idea of calculating the diagonal of a face, $\sqrt{a^2 + b^2}$, but then he gets stuck. The teacher may help him by asking: *Did you use all the data?* The student can scarcely fail to observe that the expression $\sqrt{a^2 + b^2}$

does not contain the third datum c. Therefore, he should try to bring c into play. Thus, he has a good chance to observe the decisive right triangle whose legs are $\sqrt{a^2+b^2}$ and c, and whose hypotenuse is the desired diagonal of the parallelepiped. (For another illustration see AUXILIARY ELEMENTS, 3.)

The questions we discuss here are very important. Their use in constructing the solution is clearly shown by the foregoing example. They may help us to find the weak spot in our conception of the problem. They may point out a missing element. When we know that a certain element is still missing, we naturally try to bring it into play. Thus, we have a clue, we have a definite line of inquiry to follow, and have a good chance to meet with the decisive idea.

2. The questions we discussed are helpful not only in constructing an argument but also in checking it. In order to be more concrete, let us assume that we have to check the proof of a theorem whose hypothesis consists of three parts, all three essential to the truth of the theorem. That is, if we discard any part of the hypothesis, the theorem ceases to be true. Therefore, if the proof neglects to use any part of the hypothesis, the proof must be wrong. Does the proof *use the whole hypothesis?* Does it use the first part of the hypothesis? Where does it use the first part of the hypothesis? Where does it use the second part? Where the third? Answering to all these questions we check the proof.

This sort of checking is effective, instructive, and almost necessary for thorough understanding if the argument is long and heavy—as THE INTELLIGENT READER should know.

3. The questions we discussed aim at examining the completeness of our conception of the problem. Our conception is certainly incomplete if we fail to take into

account any essential datum or condition or hypothesis. But it is also incomplete if we fail to realize the meaning of some essential term. Therefore, in order to examine our conception, we should also ask: *Have you taken into account all essential notions involved in the problem?* See DEFINITION, 7.

4. The foregoing remarks, however, are subject to caution and certain limitations. In fact, their straightforward application is restricted to problems which are "perfectly stated" and "reasonable."

A perfectly stated and reasonable "problem to find" must have all necessary data and not a single superfluous datum; also its condition must be just sufficient, neither contradictory nor redundant. In solving such a problem, we have to use, of course, all the data and the whole condition.

The object of a "problem to prove" is a mathematical theorem. If the problem is perfectly stated and reasonable, each clause in the hypothesis of the theorem must be essential to the conclusion. In proving such a theorem we have to use, of course, each clause of the hypothesis.

Mathematical problems proposed in traditional textbooks are supposed to be perfectly stated and reasonable. We should however not rely too much on this; when there is the slightest doubt, we should ask: IS IT POSSIBLE TO SATISFY THE CONDITION? Trying to answer this question, or a similar one, we may convince ourselves, at least to a certain extent, that our problem is as good as it is supposed to be.

The question stated in the title of the present article and allied questions may and should be asked without modification only when we know that the problem before us is reasonable and perfectly stated or when, at least, we have no reason to suspect the contrary.

5. There are some nonmathematical problems which

may be, in a certain sense, "perfectly stated." For instance, good chess problems are supposed to have but one solution and no superfluous piece on the chessboard, etc.

PRACTICAL PROBLEMS however are usually far from being perfectly stated and require a thorough reconsideration of the questions discussed in the present article.

Do you know a related problem? We can scarcely imagine a problem absolutely new, unlike and unrelated to any formerly solved problem; but, if such a problem could exist, it would be insoluble. In fact, when solving a problem, we always profit from previously solved problems, using their result, or their method, or the experience we acquired solving them. And, of course, the problems from which we profit must be in some way related to our present problem. Hence the question: *Do you know a related problem?*

There is usually no difficulty at all in recalling formerly solved problems which are more or less related to our present one. On the contrary, we may find too many such problems and there may be difficulty in choosing a useful one. We have to look around for closely related problems; we LOOK AT THE UNKNOWN, or we look for a formerly solved problem which is linked to our present one by GENERALIZATION, SPECIALIZATION, or ANALOGY.

The question we discuss here aims at the mobilization of our formerly acquired knowledge (PROGRESS AND ACHIEVEMENT, 1). An essential part of our mathematical knowledge is stored in the form of formerly proved theorems. Hence the question: *Do you know a theorem that could be useful?* This question may be particularly suitable when our problem is a "problem to prove," that is, when we have to prove or disprove a proposed theorem.

Draw a figure; see FIGURES. *Introduce suitable notation;* see NOTATION.

Examine your guess. Your guess may be right, but it is foolish to accept a vivid guess as a proven truth—as primitive people often do. Your guess may be wrong. But it is also foolish to disregard a vivid guess altogether—as pedantic people sometimes do. Guesses of a certain kind deserve to be examined and taken seriously: those which occur to us after we have attentively considered and really understood a problem in which we are genuinely interested. Such guesses usually contain at least a fragment of the truth although, of course, they very seldom show the whole truth. Yet there is a chance to extract the whole truth if we examine such a guess appropriately.

Many a guess has turned out to be wrong but nevertheless useful in leading to a better one.

No idea is really bad, unless we are uncritical. What is really bad is to have no idea at all.

1. *Don't.* Here is a typical story about Mr. John Jones. Mr. Jones works in an office. He had hoped for a little raise but his hope, as hopes often are, was disappointed. The salaries of some of his colleagues were raised but not his. Mr. Jones could not take it calmly. He worried and worried and finally suspected that Director Brown was responsible for his failure in getting a raise.

We cannot blame Mr. Jones for having conceived such a suspicion. There were indeed some signs pointing to Director Brown. The real mistake was that, after having conceived that suspicion, Mr. Jones became blind to all signs pointing in the opposite direction. He worried himself into firmly believing that Director Brown was his personal enemy and behaved so stupidly that he almost succeeded in making a real enemy of the director.

The trouble with Mr. John Jones is that he behaves

like most of us. He never changes his major opinions. He changes his minor opinions not infrequently and quite suddenly; but he never doubts any of his opinions, major or minor, as long as he has them. He never doubts them, or questions them, or examines them critically—he would especially hate critical examination, if he understood what that meant.

Let us concede that Mr. John Jones is right to a certain extent. He is a busy man; he has his duties at the office and at home. He has little time for doubt or examination. At best, he could examine only a few of his convictions and why should he doubt one if he has no time to examine that doubt?

Still, don't do as Mr. John Jones does. Don't let your suspicion, or guess, or conjecture, grow without examination till it becomes ineradicable. At any rate, in theoretical matters, the best of ideas is hurt by uncritical acceptance and thrives on critical examination.

2. *A mathematical example.* Of all quadrilaterals with given perimeter, find the one that has the greatest area.

What is the unknown? A quadrilateral.

What are the data? The perimeter of the quadrilateral is given.

What is the condition? The required quadrilateral should have a greater area than any other quadrilateral with the same perimeter.

This problem is very different from the usual problems in elementary geometry and, therefore, it is quite natural to start guessing.

Which quadrilateral is likely to be the one with the greatest area? What would be the simplest guess? We may have heard that of all figures with the same perimeter the circle has the greatest area; we may even suspect some reason for the plausibility of this statement. Now, which quadrilateral comes nearest to the circle? Which one comes nearest to it in symmetry?

The square is a pretty obvious guess. If we take this guess seriously, we should realize what it means. We should have the courage to state it: "Of all quadrilaterals with given perimeter the square has the greatest area." If we decide ourselves to examine this statement, the situation changes. Originally, we had a "problem to find." After having formulated our guess, we have a "problem to prove"; we have to prove or disprove the theorem formulated.

If we do not know any problem similar to ours that has been solved before, we may find our task pretty tough. *If you cannot solve the proposed problem, try to solve first some related problem. Could you solve a part of the problem?* It may occur to us that if the square is privileged among quadrilaterals it must, by that very fact, also be privileged among rectangles. A part of our problem would be solved if we could succeed in proving the following statement: "Of all rectangles with given perimeter the square has the greatest area."

This theorem appears more accessible than the former; it is, of course, weaker. At any rate, we should realize what it means; we should have the courage to restate it in more detail. We can restate it advantageously in the language of algebra.

The area of a rectangle with adjacent sides a and b is ab. Its perimeter is $2a + 2b$.

One side of the square that has the same perimeter as the rectangle just mentioned is $\frac{a+b}{2}$. Thus, the area of this square is $\left(\frac{a+b}{2}\right)^2$. It should be greater than the area of the rectangle, and so we should have

$$\left(\frac{a+b}{2}\right)^2 > ab.$$

Is this true? The same assertion can be written in the

equivalent form

$$a^2 + 2ab + b^2 > 4ab.$$

This, however, is true, for it is equivalent to

$$a^2 - 2ab + b^2 > 0$$

or to

$$(a - b)^2 > 0$$

and this inequality certainly holds, unless $a = b$, that is, the rectangle examined is a square.

We have not solved our problem yet, but we have made some progress just by facing squarely our rather obvious guesses.

3. *A nonmathematical example.* In a certain crossword puzzle we have to find a word with seven letters, and the clue is: "Do the walls again, back and forth."[4]

What is the unknown? A word.

What are the data? The length of the word is given; it has seven letters.

What is the condition? It is stated in the clue. It has something to do with walls, yet it is still very hazy.

Thus, we have to reexamine the clue. As we do so, the last part may catch our attention: ". . . again, back and forth." *Could you solve a part of the problem?* Here is a chance to guess the beginning of the word. Since the repetition is so strongly emphasized, the word, quite possibly, might start with "re." This is a pretty obvious guess. If we are tempted to believe it, we should realize what it means. The word required would look thus:

R E - - - - -

Can you check the result? If another word of the puzzle crosses the one just considered in the first letter, we have an R to start that other word. It may be a good idea

[4] *The Nation*, June 9, 1945, Crossword Puzzle, No. 119.

to switch to that other word and check the R. If we succeed in verifying that R or if, at least, we do not find any reason against it, we come back to our original word. We ask again: *What is the condition?* As we reexamine the clue, the very last part may catch our attention: ". . . back and forth." Could this imply that the word we seek can be read not only forward but backward? This is a less obvious guess (yet there are such cases, see DECOMPOSING AND RECOMBINING, 8).

At any rate, let us face this guess; let us realize what it means. The word would look as follows:

RE - - - ER.

Moreover, the third letter should be the same as the fifth; it is very likely a consonant and the fourth or middle letter a vowel.

The reader can now easily guess the word by himself. If nothing else helps, he can try all the vowels, one after the other, for the letter in the middle.

Figures are not only the object of geometric problems but also an important help for all sorts of problems in which there is nothing geometric at the outset. Thus, we have two good reasons to consider the role of figures in solving problems.

1. If our problem is a problem of geometry, we have to consider a figure. This figure may be in our imagination, or it may be traced on paper. On certain occasions, it might be desirable to imagine the figure without drawing it; but if we have to examine various details, one detail after the other, it is desirable to *draw a figure*. If there are many details, we cannot imagine all of them simultaneously, but they are all together on the paper. A detail pictured in our imagination may be forgotten; but the detail traced on paper remains, and, when we

come back to it, it reminds us of our previous remarks, it saves us some of the trouble we have in recollecting our previous consideration.

2. We now consider more specially the use of figures in problems of geometric construction.

We start the detailed consideration of such a problem by drawing a figure containing the unknown and the data, all these elements being assembled as it is prescribed by the condition of the problem. In order to understand the problem distinctly, we have to consider each datum and each part of the condition separately; then we reunite all parts and consider the condition as a whole, trying to see simultaneously the various connections required by the problem. We would scarcely be able to handle and separate and recombine all these details without a figure on paper.

On the other hand, before we have solved the problem definitively, it remains doubtful whether such a figure can be drawn at all. Is it possible to satisfy the whole condition imposed by the problem? We are not entitled to say Yes before we have obtained the definitive solution; nevertheless we begin with assuming a figure in which the unknown is connected with the data as prescribed by the condition. It seems that, drawing the figure, we have made an unwarranted assumption.

No, we have not. Not necessarily. We do not act incorrectly when, examining our problem, we consider the *possibility* that there is an object that satisfies the condition imposed upon the unknown and has, with all the data, the required relations, provided we do not confuse mere possibility with certainty. A judge does not act incorrectly when, questioning the defendant, he considers the hypothesis that the defendant perpetrated the crime in question, provided he does not commit himself to this hypothesis. Both the mathematician and the judge may examine a possibility without prejudice, postponing their judgment till the examination yields some definite result.

The method of starting the examination of a problem of construction by drawing a sketch on which, supposedly, the condition is satisfied, goes back to the Greek geometers. It is hinted by the short, somewhat enigmatic phrase of Pappus: *Assume what is required to be done as already done.* The following recommendation is somewhat less terse but clearer: *Draw a hypothetical figure which supposes the condition of the problem satisfied in all its parts.*

This is a recommendation for problems of geometric construction but in fact there is no need to restrict us to any such particular kind of problem. We may extend the recommendation to all "problems to find" stating it in the following general form: *Examine the hypothetical situation in which the condition of the problem is supposed to be fully satisfied.*

Compare PAPPUS, 6.

3. Let us discuss a few points about the actual drawing of figures.

(I) Shall we draw the figures exactly or approximately, with instruments or free-hand?

Both kinds of figures have their advantages. Exact figures have, in principle, the same role in geometry as exact measurements in physics; but, in practice, exact figures are less important than exact measurements because the theorems of geometry are much more extensively verified than the laws of physics. The beginner, however, should construct many figures as exactly as he can in order to acquire a good experimental basis; and exact figures may suggest geometric theorems also to the more advanced. Yet, for the purpose of reasoning, carefully drawn free-hand figures are usually good enough, and they are much more quickly done. Of course, the figure should not look absurd; lines supposed to be straight should not be wavy, and so-called circles should not look like potatoes.

An inaccurate figure can occasionally suggest a false conclusion, but the danger is not great and we can protect ourselves from it by various means, especially by varying the figure. There is no danger if we concentrate upon the logical connections and realize that the figure is a help, but by no means the basis of our conclusions; the logical connections constitute the real basis. [This point is instructively illustrated by certain well known paradoxes which exploit cleverly the intentional inaccuracy of the figure.]

(II) It is important that the elements are assembled in the required relations, it is unimportant in which order they are constructed. Therefore, choose the most convenient order. For example, to illustrate the idea of trisection, you wish to draw two angles, α and β, so that $\alpha = 3\beta$. Starting from an arbitrary α, you cannot construct β with ruler and compasses. Therefore, you choose a fairly small, but otherwise arbitrary β and, starting from β, you construct α which is easy.

(III) Your figure should not suggest any undue specialization. The different parts of the figure should not exhibit apparent relations not required by the problem. Lines should not seem to be equal, or to be perpendicular, when they are not necessarily so. Triangles should not seem to be isosceles, or right-angled, when no such property is required by the problem. The triangle having the angles 45°, 60°, 75° is the one which, in a precise sense of the word, is the most "remote" both from the isosceles, and from the right-angled shape.[5] You draw

[5] If the angles of a triangle are α, β, γ, and $90° > \alpha > \beta > \gamma$, then at least one of the differences $90° - \alpha$, $\alpha - \beta$, $\beta - \gamma$ is $< 15°$, unless $\alpha = 75°$, $\beta = 60°$, $\gamma = 45°$. In fact,

$$\frac{3(90° - \alpha) + 2(\alpha - \beta) + (\beta - \gamma)}{6} = 15°.$$

this, or a not very different triangle, if you wish to consider a "general" triangle.

(IV) In order to emphasize the different roles of different lines, you may use heavy and light lines, continuous and dotted lines, or lines in different colors. You draw a line very lightly if you are not yet quite decided to use it as an auxiliary line. You may draw the given elements with red pencil, and use other colors to emphasize important parts, as a pair of similar triangles, etc.

(V) In order to illustrate solid geometry, shall we use three-dimensional models, or drawings on paper and blackboard?

Three-dimensional models are desirable, but troublesome to make and expensive to buy. Thus, usually, we must be satisfied with drawings although it is not easy to make them impressive. Some experimentation with self-made cardboard models is very desirable for beginners. It is helpful to take objects of our everyday surroundings as representations of geometric notions. Thus, a box, a tile, or the classroom may represent a rectangular parallelepiped, a pencil, a circular cylinder, a lampshade, the frustum of a right circular cone, etc.

4. Figures traced on paper are easy to produce, easy to recognize, easy to remember. Plane figures are especially familiar to us, problems about plane figures especially accessible. We may take advantage of this circumstance, we may use our aptitude for handling figures in handling nongeometrical objects if we contrive to find a suitable geometrical representation for those nongeometrical objects.

In fact, geometrical representations, graphs and diagrams of all sorts, are used in all sciences, not only in physics, chemistry, and the natural sciences, but also in economics, and even in psychology. Using some suitable geometrical representation, we try to express everything in the language of figures, to reduce all sorts of problems to problems of geometry.

Thus, even if your problem is not a problem of geometry, you may try to *draw a figure.* To find a lucid geometric representation for your nongeometrical problem could be an important step toward the solution.

Generalization is passing from the consideration of one object to the consideration of a set containing that object; or passing from the consideration of a restricted set to that of a more comprehensive set containing the restricted one.

1. If, by some chance, we come across the sum

$$1 + 8 + 27 + 64 = 100$$

we may observe that it can be expressed in the curious form

$$1^3 + 2^3 + 3^3 + 4^3 = 10^2.$$

Now, it is natural to ask ourselves: Does it often happen that a sum of successive cubes as

$$1^3 + 2^3 + 3^3 + \cdots + n^3$$

is a square? In asking this, we generalize. This generalization is a lucky one; it leads from one observation to a remarkable general law. Many results were found by lucky generalizations in mathematics, physics, and the natural sciences. See INDUCTION AND MATHEMATICAL INDUCTION.

2. Generalization may be useful in the solution of problems. Consider the following problem of solid geometry: "A straight line and a regular octahedron are given in position. Find a plane that passes through the given line and bisects the volume of the given octahedron." This problem may look difficult but, in fact, very little

familiarity with the shape of the regular octahedron is sufficient to suggest the following more general problem: "A straight line and a *solid with a center of symmetry* are given in position. Find a plane that passes through the given line and bisects the volume of the given solid." The plane required passes, of course, through the center of symmetry of the solid, and is determined by this point and the given line. As the octahedron has a center of symmetry, our original problem is also solved.

The reader will not fail to observe that the second problem is more general than the first, and, nevertheless, much easier than the first. In fact, our main achievement in solving the first problem was to *invent the second problem*. Inventing the second problem, we recognize the role of the center of symmetry; we *disentangled* that property of the octahedron which is essential for the problem at hand, namely that it has such a center.

The more general problem may be easier to solve. This sounds paradoxical but, after the foregoing example, it should not be paradoxical to us. The main achievement in solving the special problem was to invent the general problem. After the main achievement, only a minor part of the work remains. Thus, in our case, the solution of the general problem is only a minor part of the solution of the special problem.

See INVENTOR'S PARADOX.

3. "Find the volume of the frustum of a pyramid with square base, being given that the side of the lower base is 10 in., the side of the upper base 5 in., and the altitude of the frustum 6 in." If for the numbers 10, 5, 6 we substitute letters, for instance a, b, h, we generalize. We obtain a more general problem than the original one, namely the following: "Find the volume of the frustum of a pyramid with square base, being given that the side of the lower base is a, the side of the upper base b, and

the altitude of the frustum h." Such generalization may be very useful. Passing from a problem "in numbers" to another one "in letters" we gain access to new procedures; we can vary the data, and, doing so, we may check our results in various ways. See CAN YOU CHECK THE RESULT? 2, VARIATION OF THE PROBLEM, 4.

Have you seen it before? It is possible that we have solved before the same problem that we have to do now, or that we have heard of it, or that we had a very similar problem. These are possibilities which we should not fail to explore. We try to remember what happened. *Have you seen it before? Or have you seen the same problem in a slightly different form?* Even if the answer is negative such questions may start the mobilization of useful knowledge.

The question in the title of the present article is often used in a more general meaning. In order to obtain the solution, we have to extract relevant elements from our memory, we have to mobilize the pertinent parts of our dormant knowledge (PROGRESS AND ACHIEVEMENT). We cannot know, of course, in advance which parts of our knowledge may be relevant; but there are certain possibilities which we should not fail to explore. Thus, any feature of the present problem that played a role in the solution of some other problem may play again a role. Therefore, if any feature of the present problem strikes us as possibly important, we try to recognize it. What is it? Is it familiar to you? *Have you seen it before?*

Here is a problem related to yours and solved before. This is good news; a problem for which the solution is known and which is connected with our present problem, is certainly welcome. It is still more welcome if the connection is close and the solution simple. There is a good chance that such a problem will be useful in solving our present one.

The situation that we are discussing here is typical and important. In order to see it clearly let us compare it with the situation in which we find ourselves when we are working at an auxiliary problem. In both cases, our aim is to solve a certain problem A and we introduce and consider another problem B in the hope that we may derive some profit for the solution of the proposed problem A from the consideration of that other problem B. The difference is in our relation to B. Here, we succeeded in recollecting an old problem B of which we know the solution but we do not know yet how to use it. There, we succeeded in inventing a new problem B; we know (or at least we suspect strongly) how to use B, but we do not know yet how to solve it. Our difficulty concerning B makes all the difference between the two situations. When this difficulty is overcome, we may use B in the same way in both cases; we may use the result or the method (as explained in AUXILIARY PROBLEM, 3), and, if we are lucky, we may use both the result and the method. In the situation considered here, we know well the solution of B but we do not know yet how to use it. Therefore, we ask: *Could you use it? Could you use its result? Could you use its method?*

The intention of using a certain formerly solved problem influences our conception of the present problem. Trying to link up the two problems, the new and the old, we introduce into the new problem elements corresponding to certain important elements of the old problem. For example, our problem is to determine the sphere circumscribed about a given tetrahedron. This is a problem of solid geometry. We may remember that we have solved before the analogous problem of plane geometry of constructing the circle circumscribed about a given triangle. Then we recollect that in the old problem of plane geometry, we used the perpendicular bisectors of the sides of the triangle. It is reasonable to try to introduce something analogous into our present problem. Thus we may be led to introduce into our present problem, as corresponding auxiliary elements, the perpendicular bisecting planes of the edges of the tetrahedron. After this idea, we can easily work out the solution to the problem of solid geometry, following the analogous solution in plane geometry.

The foregoing example is typical. The consideration of a formerly solved related problem leads us to the introduction of auxiliary elements, and the introduction of suitable auxiliary elements makes it possible for us to use the related problem to full advantage in solving our present problem. We aim at such an effect when, thinking about the possible use of a formerly solved related problem, we ask: *Should you introduce some auxiliary element in order to make its use possible?*

Here is a theorem related to yours and proved before. This version of the remark discussed here is exemplified in section **19.**

Heuristic, or heuretic, or "ars inveniendi" was the name of a certain branch of study, not very clearly circumscribed, belonging to logic, or to philosophy, or to psychology, often outlined, seldom presented in detail, and as good as forgotten today. The aim of heuristic is to study the methods and rules of discovery and invention. A few traces of such study may be found in the commentators of Euclid; a passage of PAPPUS is particularly interesting in this respect. The most famous attempts to build up a system of heuristic are due to DESCARTES and to LEIBNITZ, both great mathematicians and philosophers. Bernard BOLZANO presented a notable detailed account of heuristic. The present booklet is an attempt to revive heuristic in a modern and modest form. See MODERN HEURISTIC.

Heuristic, as an adjective, means "serving to discover."

Heuristic reasoning is reasoning not regarded as final and strict but as provisional and plausible only, whose purpose is to discover the solution of the present problem. We are often obliged to use heuristic reasoning. We shall attain complete certainty when we shall have obtained the complete solution, but before obtaining certainty we must often be satisfied with a more or less plausible guess. We may need the provisional before we attain the final. We need heuristic reasoning when we construct a strict proof as we need scaffolding when we erect a building.

See SIGNS OF PROGRESS. Heuristic reasoning is often based on induction, or on analogy; see INDUCTION AND MATHEMATICAL INDUCTION, and ANALOGY, 8, 9, 10.[6]

Heuristic reasoning is good in itself. What is bad is to mix up heuristic reasoning with rigorous proof. What is worse is to sell heuristic reasoning for rigorous proof.

The teaching of certain subjects, especially the teaching of calculus to engineers and physicists, could be essentially improved if the nature of heuristic reasoning were better understood, both its advantages and its limitations openly recognized, and if the textbooks would present heuristic arguments openly. A heuristic argument presented with taste and frankness may be useful; it may prepare for the rigorous argument of which it usually contains certain germs. But a heuristic argument is likely to be harmful if it is presented ambiguously with visible

[6] See also a paper by the author in American Mathematical Monthly, vol. 48, pp. 450-465.

hesitation between shame and pretension. See WHY PROOFS?

If you cannot solve the proposed problem do not let this failure afflict you too much but try to find consolation with some easier success, *try to solve first some related problem;* then you may find courage to attack your original problem again. Do not forget that human superiority consists in going around an obstacle that cannot be overcome directly, in devising some suitable auxiliary problem when the original one appears insoluble.

Could you imagine a more accessible related problem? You should now *invent* a related problem, not merely *remember* one; I hope that you have tried already the question: *Do you know a related problem?*

The remaining questions in that paragraph of the list which starts with the title of the present article have a common aim, the VARIATION OF THE PROBLEM. There are different means to attain this aim as GENERALIZATION, SPECIALIZATION, ANALOGY, and others which are various ways of DECOMPOSING AND RECOMBINING.

Induction and mathematical induction. Induction is the process of discovering general laws by the observation and combination of particular instances. It is used in all sciences, even in mathematics. Mathematical induction is used in mathematics alone to prove theorems of a certain kind. It is rather unfortunate that the names are connected because there is very little logical connection between the two processes. There is, however, some practical connection; we often use both methods together. We are going to illustrate both methods by the same example.

1. We may observe, by chance, that

$$1 + 8 + 27 + 64 = 100$$

and, recognizing the cubes and the square, we may give to the fact we observed the more interesting form:

$$1^3 + 2^3 + 3^3 + 4^3 = 10^2.$$

How does such a thing happen? Does it often happen that such a sum of successive cubes is a square?

In asking this we are like the naturalist who, impressed by a curious plant or a curious geological formation, conceives a general question. Our general question is concerned with the sum of successive cubes

$$1^3 + 2^3 + 3^3 + \cdots + n^3.$$

We were led to it by the "particular instance" $n = 4$.

What can we do for our question? What the naturalist would do; we can investigate other special cases. The special cases $n = 2, 3$ are still simpler, the case $n = 5$ is the next one. Let us add, for the sake of uniformity and completeness, the case $n = 1$. Arranging neatly all these cases, as a geologist would arrange his specimens of a certain ore, we obtain the following table:

$$\begin{aligned}
1 \qquad\qquad\qquad\qquad\qquad &= 1 = 1^2 \\
1 + 8 \qquad\qquad\qquad\qquad &= 9 = 3^2 \\
1 + 8 + 27 \qquad\qquad\qquad &= 36 = 6^2 \\
1 + 8 + 27 + 64 \qquad\quad &= 100 = 10^2 \\
1 + 8 + 27 + 64 + 125 &= 225 = 15^2.
\end{aligned}$$

It is hard to believe that all these sums of consecutive cubes are squares by mere chance. In a similar case, the naturalist would have little doubt that the general law suggested by the special cases heretofore observed is correct; the general law is almost proved by *induction*. The mathematician expresses himself with more reserve although fundamentally, of course, he thinks in the same fashion. He would say that the following theorem is strongly suggested by induction:

The sum of the first n *cubes is a square.*

2. We have been led to conjecture a remarkable, somewhat mysterious law. Why should those sums of successive cubes be squares? But, apparently, they are squares.

What would the naturalist do in such a situation? He would go on examining his conjecture. In so doing, he may follow various lines of investigation. The naturalist may accumulate further experimental evidence; if we wish to do the same, we have to test the next cases, $n = 6, 7, \dots$. The naturalist may also reexamine the facts whose observation has led him to his conjecture; he compares them carefully, he tries to disentangle some deeper regularity, some further analogy. Let us follow this line of investigation.

Let us reexamine the cases $n = 1, 2, 3, 4, 5$ which we arranged in our table. Why are all these sums squares? What can we say about these squares? Their bases are 1, 3, 6, 10, 15. What about these bases? Is there some deeper regularity, some further analogy? At any rate, they do not seem to increase too irregularly. How do they increase? The difference between two successive terms of this sequence is itself increasing,

$$3 - 1 = 2, \quad 6 - 3 = 3, \quad 10 - 6 = 4, \quad 15 - 10 = 5.$$

Now these differences are conspicuously regular. We may see here a surprising analogy between the bases of those squares, we may see a remarkable regularity in the numbers 1, 3, 6, 10, 15:

$$\begin{aligned} 1 &= 1 \\ 3 &= 1 + 2 \\ 6 &= 1 + 2 + 3 \\ 10 &= 1 + 2 + 3 + 4 \\ 15 &= 1 + 2 + 3 + 4 + 5. \end{aligned}$$

If this regularity is general (and the contrary is hard to

believe) the theorem we suspected takes a more precise form:

It is, for n = 1, 2, 3, . . .

$$1^3 + 2^3 + 3^3 + \cdots + n^3 = (1 + 2 + 3 + \cdots + n)^2.$$

3. The law we just stated was found by induction, and the manner in which it was found conveys to us an idea about induction which is necessarily one-sided and imperfect but not distorted. Induction tries to find regularity and coherence behind the observations. Its most conspicuous instruments are generalization, specialization, analogy. Tentative generalization starts from an effort to understand the observed facts; it is based on analogy, and tested by further special cases.

We refrain from further remarks on the subject of induction about which there is wide disagreement among philosophers. But it should be added that many mathematical results were found by induction first and proved later. Mathematics presented with rigor is a systematic deductive science but mathematics in the making is an experimental inductive science.

4. In mathematics as in the physical sciences we may use observation and induction to discover general laws. But there is a difference. In the physical sciences, there is no higher authority than observation and induction but in mathematics there is such an authority: rigorous proof.

After having worked a while experimentally it may be good to change our point of view. Let us be strict. We have discovered an interesting result but the reasoning that led to it was merely plausible, experimental, provisional, heuristic; let us try to establish it definitively by a rigorous proof.

We have arrived now at a "problem to prove": to

prove or to disprove the result stated before (see 2, above).

There is a minor simplification. We may know that

$$1 + 2 + 3 + \cdots + n = \frac{n(n+1)}{2}.$$

At any rate, this is easy to verify. Take a rectangle with sides n and $n + 1$, and divide it in two halves by a zigzag line as in Fig. 15a which shows the case $n = 4$. Each of the halves is "staircase-shaped" and its area has the expression $1 + 2 + \cdots + n$; for $n = 4$ it is $1 + 2 + 3 + 4$, see Fig. 18b. Now, the whole area of the rectangle is $n(n + 1)$ of which the staircase-shaped area is one half; this proves the formula.

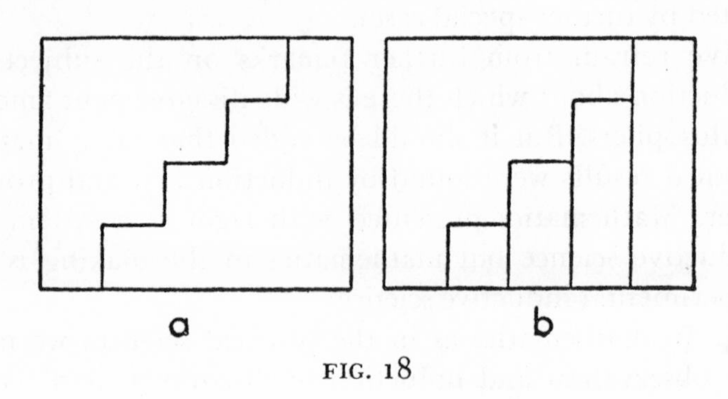

FIG. 18

We may transform the result which we found by induction into

$$1^3 + 2^3 + 3^3 + \cdots + n^3 = \left(\frac{n(n+1)}{2}\right)^2$$

5. If we have no idea how to prove this result, we may at least test it. Let us test the first case we have not tested yet, the case $n = 6$. For this value, the formula yields

$$1 + 8 + 27 + 64 + 125 + 216 = \left(\frac{6 \times 7}{2}\right)^2$$

and, on computation, this turns out to be true, both sides being equal to 441.

We can test the formula more effectively. The formula is, very likely, generally true, true for all values of n. Does it remain true when we pass from any value n to the next value $n + 1$? Along with the formula as written above (p. 118) we should also have

$$1^3 + 2^3 + 3^3 + \quad + n^3 + (n+1)^3 = \left(\frac{(n+1)(n+2)}{2}\right)^2.$$

Now, there is a simple check. Subtracting from this the formula written above, we obtain

$$(n+1)^3 = \left(\frac{(n+1)(n+2)}{2}\right)^2 - \left(\frac{n(n+1)}{2}\right)^2.$$

This is, however, easy to check. The right hand side may be written as

$$\left(\frac{n+1}{2}\right)^2 [(n+2)^2 - n^2] = \left(\frac{n+1}{2}\right)^2 [n^2 + 4n + 4 - n^2]$$

$$\frac{(n+1)^2}{4}(4n+4) = (n+1)^2(n+1) = (n+1)^3.$$

Our experimentally found formula passed a vital test.

Let us see clearly what this test means. We verified beyond doubt that

$$(n+1)^3 = \left(\frac{(n+1)(n+2)}{2}\right)^2 - \left(\frac{n(n+1)}{2}\right)^2.$$

We do not know yet whether

$$1^3 + 2^3 + 3^3 + \cdots + n^3 = \left(\frac{n(n+1)}{2}\right)^2$$

is true. But *if* we knew that this *was* true we could infer, by adding the equation which we verified beyond doubt, that

$$1^3 + 2^3 + 3^3 + \cdots + n^3 + (n+1)^3 = \left(\frac{(n+1)(n+2)}{2}\right)^2$$

is *also* true which is the same assertion for the next integer $n + 1$. Now, we actually know that our conjecture is true for $n = 1, 2, 3, 4, 5, 6$. By virtue of what we have just said, the conjecture, being true for $n = 6$, must also be true for $n = 7$; being true for $n = 7$ it is true for $n = 8$; being true for $n = 8$ it is true for $n = 9$; and so on. It holds for all n, it is proved to be true generally.

6. The foregoing proof may serve as a pattern in many similar cases. What are the essential lines of this pattern?

The assertion we have to prove must be given in advance, in precise form.

The assertion must depend on an integer n.

The assertion must be sufficiently "explicit" so that we have some possibility of testing whether it remains true in the passage from n to the next integer $n + 1$.

If we succeed in testing this effectively, we may be able to use our experience, gained in the process of testing, to conclude that the assertion must be true for $n + 1$ provided it is true for n. When we are so far it is sufficient to know that the assertion is true for $n = 1$; hence it follows for $n = 2$; hence it follows for $n = 3$, and so on; passing from any integer to the next, we prove the assertion generally.

This process is so often used that it deserves a name. We could call it "proof from n to $n + 1$" or still simpler "passage to the next integer." Unfortunately, the accepted technical term is "mathematical induction." This name results from a random circumstance. The precise assertion that we have to prove may come from any source, and it is immaterial from the logical viewpoint what the source is. Now, in many cases, as in the case we discussed here in detail, the source is induction, the assertion is found experimentally, and so the proof appears as a mathematical complement to induction; this explains the name.

7. Here is another point, somewhat subtle, but important to anybody who desires to find proofs by himself. In the foregoing, we found two different assertions by observation and induction, one after the other, the first under 1, the second under 2; the second was more precise than the first. Dealing with the second assertion, we found a possibility of checking the passage from n to $n + 1$, and so we were able to find a proof by "mathematical induction." Dealing with the first assertion, and ignoring the precision added to it by the second one, we should scarcely have been able to find such a proof. In fact, the first assertion is less precise, less "explicit," less "tangible," less accessible to testing and checking than the second one. Passing from the first to the second, from the less precise to the more precise statement, was an important preparative for the final proof.

This circumstance has a paradoxical aspect. The second assertion is stronger; it implies immediately the first, whereas the somewhat "hazy" first assertion can hardly imply the more "clear-cut" second one. Thus, the stronger theorem is easier to master than the weaker one; this is the INVENTOR'S PARADOX.

Inventor's paradox. The more ambitious plan may have more chances of success.

This sounds paradoxical. Yet, when passing from one problem to another, we may often observe that the new, more ambitious problem is easier to handle than the original problem. More questions may be easier to answer than just one question. The more comprehensive theorem may be easier to prove, the more general problem may be easier to solve.

The paradox disappears if we look closer at a few examples (GENERALIZATION, 2; INDUCTION AND MATHEMATICAL INDUCTION, 7). The more ambitious plan may have more chances of success provided it is not based on mere pretension but on some vision of the things beyond those immediately present.

Is it possible to satisfy the condition? *Is the condition sufficient to determine the unknown? Or is it insufficient? Or redundant? Or contradictory?*

These questions are often useful at an early stage when they do not need a final answer but just a provisional answer, a guess. For examples, see sections **8, 18.**

It is good to foresee any feature of the result for which we work. When we have some idea of what we can expect, we know better in which direction we should go. Now, an important feature of a problem is the number of solutions of which it admits. Most interesting among problems are those which admit of just one solution; we are inclined to consider problems with a uniquely determined solution as the only "reasonable" problems. Is our problem, in this sense, "reasonable"? If we can answer this question, even by a plausible guess, our interest in the problem increases and we can work better.

Is our problem "reasonable"? This question is useful at an early stage of our work *if* we can answer it easily. If the answer is difficult to obtain, the trouble we have in obtaining it may outweigh the gain in interest. The same is true of the question "*Is it possible to satisfy the condition?*" and the allied questions of our list. We should put them because the answer might be easy and plausible, but we should not insist on them when the answer seems to be difficult or obscure.

The corresponding questions for "problems to prove" are: *Is it likely that the proposition is true? Or is it more likely that it is false?* The way the question is put shows clearly that only a guess, a plausible provisional answer, is expected.

Leibnitz, Gottfried Wilhelm (1646-1716), great mathematician and philosopher, planned to write an "Art of Invention" but he never carried through his plan. Numerous fragments dispersed in his works show, however, that he entertained interesting ideas about the subject whose importance he often emphasized. Thus, he wrote: "Nothing is more important than to see the sources of invention which are, in my opinion, more interesting than the inventions themselves."

Lemma means "auxiliary theorem." The word is of Greek origin; a more literal translation would be "what is assumed."

We are trying to prove a theorem, say, A. We are led to suspect another theorem, say, B; if B were true we could perhaps, using it, prove A. We assume B provisionally, postponing its proof, and go ahead with the proof of A. Such a theorem B is assumed, and is an auxiliary theorem to the originally proposed theorem A. Our little story is fairly typical and explains the present meaning of the word "lemma."

Look at the unknown. This is old advice; the corresponding Latin saying is: "respice finem." That is, look at the end. Remember your aim. Do not forget your goal. Think of what you are desiring to obtain. Do not lose sight of what is required. Keep in mind what you are working for. *Look at the unknown. Look at the conclusion.* The last two versions of "respice finem" are specifically adapted to mathematical problems, to "problems to find" and to "problems to prove" respectively.

Focusing our attention on our aim and concentrating our will on our purpose, we think of ways and means to attain it. What are the means to this end? How can you attain your aim? How can you obtain a result of this kind? What causes could produce such a result? Where have you seen such a result produced? What do people usually do to obtain such a result? *And try to think of a familiar problem having the same or a similar unknown. And try to think of a familiar theorem having the same or a similar conclusion.* Again, the last two versions are specifically adapted to "problems to find" and to "problems to prove" respectively.

1. We are going to consider mathematical problems, "problems to find," and the suggestion: *Try to think of a familiar problem having the same unknown.* Let us compare this suggestion with that involved in the question: *Do you know a related problem?*

The latter suggestion is more general than the former one. If a problem is related to another problem, the two have something in common; they may involve a few common objects or notions, or have some data in common, or some part of the condition, and so on. Our first suggestion insists on a particular common point: The two problems should have the same unknown. That is, the unknown should be in both cases an object of the same category, for instance, in both cases the length of a straight line.

In comparison with the general suggestion, there is a certain economy in the special suggestion.

First, we may save some effort in representing the problem; we must not look at once at the whole problem but just at the unknown. The problem appears to us schematically, as

"Given . . . . find the length of the line."

Second, there is a certain economy of choice. Many, many problems may be related to the proposed problem, having some point or other in common with it. But, looking at the unknown, we restrict our choice; we take into consideration only such problems as have the same unknown. And, of course, among the problems having the same unknown, we consider first those which are the most elementary and the most familiar to us.

2. The problem before us has the form:

"Given . . . . find the length of the line."

Now the simplest and most familiar problems of this kind are concerned with triangles: Given three constituent parts of a triangle find the length of a side. Remembering this, we have found something that may be relevant: *Here is a problem related to yours and solved before. Could you use it? Could you use its result?* In order to use the familiar results about triangles, we must have a triangle in our figure. Is there a triangle? Or should we introduce one in order to profit from those familiar results? *Should you introduce some auxiliary element in order to make their use possible?*

There are several simple problems whose unknown is the side of a triangle. (They differ from each other in the data; two angles may be given and one side, or two sides and one angle, and the position of the angle with respect to the given sides may be different. Then, all these problems are particularly simple for right triangles.) With our attention riveted upon the problem before us, we try to find out which kind of triangle we should introduce, which formerly solved problem (with the same unknown as that before us) we could most conveniently adapt to our present purpose.

Having introduced a suitable auxiliary triangle, it may happen that we do not know yet three constituent parts of it. This, however, is not absolutely necessary; if we foresee that the missing parts can be obtained somehow we have made essential progress, we have a plan of the solution.

3. The procedure sketched in the foregoing (under 1 and 2) is illustrated, essentially, by section **10** (the illustration is somewhat obscured by the slowness of the students). It is not difficult at all to add many similar examples. In fact, the solution of almost all "problems to find" usually proposed in less advanced classes can be started by proper use of the suggestion: *And try to think of a familiar problem having the same or a similar unknown.*

We must take such problems schematically, and look at the unknown first:

(1) Given . . . . find the length of the line.
(2) Given . . . . find the angle.
(3) Given . . . . find the volume of the tetrahedron.
(4) Given . . . . construct the point.

If we have some experience in dealing with elementary mathematical problems, we will readily recall some simple and familiar problem or problems having the same unknown. If the problem proposed is not one of those simple familiar problems we naturally try to make use of what is familiar to us and profit from the result of those simple problems. We try to introduce some useful well-known thing into the problem, and doing so we may get a good start.

In each of the four cases mentioned there is an obvious plan, a plausible guess about the future course of the solution.

(1) The unknown should be obtained as a side of some triangle. It remains to introduce a suitable triangle with three known, or easily obtainable, constituents.

(2) The unknown should be obtained as an angle in some triangle. It remains to introduce a suitable triangle.

(3) The unknown can be obtained if the area of the base and the length of the altitude are known. It re-

mains to find the area of a face and the corresponding altitude.

(4) The unknown should be obtained as the intersection of two loci each of which is either a circle or a straight line. It remains to disentangle such loci from the proposed condition.

In all these cases the plan is suggested by a simple problem with the same unknown and by the desire to use its result or its method. Pursuing such a plan, we may run into difficulties, of course, but we have some idea to start with which is a great advantage.

4. There is no such advantage if there is no formerly solved problem having the same unknown as the proposed problem. In such cases, it is much more difficult to tackle the proposed problem.

"Find the area of the surface of a sphere with given radius." This problem was solved by Archimedes. There is scarcely a simpler problem with the same unknown and there was certainly no such simpler problem of which Archimedes could have made use. In fact, Archimedes' solution may be regarded as one of the most notable mathematical achievements.

"Find the area of the surface of the sphere inscribed in a tetrahedron whose six edges are given." If we know Archimedes' result, we need not have Archimedes' genius to solve the problem; it remains to express the radius of the inscribed sphere in terms of the six edges of the tetrahedron. This is not exactly easy but the difficulty cannot be compared with that of Archimedes' problem.

To know or not to know a formerly solved problem with the same unknown may make all the difference between an easy and a difficult problem.

5. When Archimedes found the area of the surface of the sphere he did not know, as we just mentioned, any formerly solved problem having the same unknown. But

he knew various formerly solved problems having a similar unknown. There are curved surfaces whose area is easier to obtain than that of the sphere and which were well known in Archimedes' time, as the lateral surfaces of right circular cylinders, of right circular cones, and of the frustums of such cones. We may be certain that Archimedes considered carefully these simpler similar cases. In fact, in his solution, he uses as approximation to the sphere a composite solid consisting of two cones and several frustums of cones (see DEFINITION, 6).

If we are unable to find a formerly solved problem having the same unknown as the problem before us, we try to find one having a similar unknown. Problems of the latter kind are less closely related to the problem before us than problems of the former kind and, therefore, less easy to use for our purpose in general but they may be valuable guides nevertheless.

6. We add a few remarks concerning "problems to prove"; they are analogous to the foregoing more extensive comments on "problems to find."

We have to prove (or disprove) a clearly stated theorem. Any theorem proved in the past which is in some way related to the theorem before us has a chance to be of some service. Yet we may expect the most immediate service of theorems which have the same conclusion as the one before us. Knowing this, we *look at the conclusion,* that is, we consider our theorem emphasizing the conclusion. Our way of looking at the theorem can be expressed in writing by a scheme as:

"If . . . . then the angles are equal."

We focus our attention upon the conclusion before us and try to *think of a familiar theorem having the same or a similar conclusion.* Especially, we try to think of very simple familiar theorems of this sort.

In our case, there are various theorems of this kind and we may recollect the following: "If two triangles are congruent the corresponding angles are equal." *Here is a theorem related to yours and proved before. Could you use it? Should you introduce some auxiliary element in order to make its use possible?*

Following these suggestions, and trying to judge the help afforded by the theorem we recollected, we may conceive a plan: Let us try to prove the equality of the angles in question from congruent triangles. We see that we must introduce a pair of triangles containing those angles and prove that they are congruent. Such a plan is certainly good to start the work and it may lead eventually to the desired end as in section **19.**

7. Let us sum up. Recollecting formerly solved problems with the same or a similar unknown (formerly proved theorems with the same or a similar conclusion) we have a good chance to start in the right direction and we may conceive a plan of the solution. In simple cases, which are the most frequent in less advanced classes, the most elementary problems with the same unknown (theorems with the same conclusion) are usually sufficient. Trying to recollect problems with the same unknown is an obvious and common-sense device (compare what was said in this respect in section 4). It is rather surprising that such a simple and useful device is not more widely known; the author is inclined to think that it was not even stated before in full generality. In any case, neither students nor teachers of mathematics can afford to ignore the proper use of the suggestion: *Look at the unknown! And try to think of a familiar problem having the same or a similar unknown.*

Modern heuristic endeavors to understand the process of solving problems, especially the *mental operations* *typically useful* in this process. It has various sources of information none of which should be neglected. A serious study of heuristic should take into account both the logical and the psychological background, it should not neglect what such older writers as Pappus, Descartes, Leibnitz, and Bolzano have to say about the subject, but it should least neglect unbiased experience. Experience in solving problems and experience in watching other people solving problems must be the basis on which heuristic is built. In this study, we should not neglect any sort of problem, and should find out common features in the way of handling all sorts of problems; we should aim at general features, independent of the subject matter of the problem. The study of heuristic has "practical" aims; a better understanding of the mental operations typically useful in solving problems could exert some good influence on teaching, especially on the teaching of mathematics.

The present book is a first attempt toward the realization of this program. We are going to discuss how the various articles of this Dictionary fit into the program.

1. Our list is, in fact, a list of mental operations typically useful in solving problems; the questions and suggestions listed hint at such operations. Some of these operations are described again in the Second Part, and some of them are more thoroughly discussed and illustrated in the First Part.

For additional information about particular questions and suggestions of the list, the reader should refer to those fifteen articles of the Dictionary whose titles are the first sentences of the fifteen paragraphs of the list: WHAT IS THE UNKNOWN? IS IT POSSIBLE TO SATISFY THE CONDITION? DRAW A FIGURE. . . . CAN YOU USE THE RESULT? The reader, wishing information about a particular item of the list, should look at the first words of the paragraph in which the item is contained and then look up the article in the Dictionary that has those first words as title. For instance, the suggestion *"Go back to definitions"* is contained in the paragraph of the list whose first sentence is: COULD YOU RESTATE THE PROBLEM? Under this title, the reader finds a cross-reference to DEFINITION in which article the suggestion in question is explained and illustrated.

2. The process of solving problems is a complex process that has several different aspects. The twelve principal articles of this Dictionary study certain of these aspects at some length; we are going to mention their titles in what follows.

When we are working intensively, we feel keenly the progress of our work; we are elated when our progress is rapid, we are depressed when it is slow. What is essential to PROGRESS AND ACHIEVEMENT in solving problems? The article discussing this question is often quoted in other parts of the Dictionary and should be read fairly early.

Trying to solve a problem, we consider different aspects of it in turn, we roll it over and over incessantly in our mind; VARIATION OF THE PROBLEM is essential to our work. We may vary the problem by DECOMPOSING AND RECOMBINING its elements, or by going back to the DEFINITION of certain of its terms, or we may use the great resources of GENERALIZATION, SPECIALIZATION, and ANALOGY. Variation of the problem may lead us to AUXILIARY ELEMENTS, or to the discovery of a more accessible AUXILIARY PROBLEM.

We have to distinguish carefully between two kinds of problems, PROBLEMS TO FIND, PROBLEMS TO PROVE. Our list is specially adapted to "problems to find." We have to revise it and change some of its questions and suggestions in order to apply it also to "problems to prove."

In all sorts of problems, but especially in mathematical problems which are not too simple, suitable NOTATION and geometrical FIGURES are a great and often indispensable help.

3. The process of solving problems has many aspects but some of them are not considered at all in this book and others only very briefly. It is justified, I think, to exclude from a first short exposition points which could appear too subtle, or too technical, or too controversial.

Provisional, merely plausible HEURISTIC REASONING is important in discovering the solution, but you should not take it for a proof; you must guess, but also EXAMINE YOUR GUESS. The nature of heuristic arguments is discussed in SIGNS OF PROGRESS, but the discussion could go further.

The consideration of certain logical patterns is important in our subject but it appeared advisable not to introduce any technical article. There are only two articles predominantly devoted to psychological aspects, on DETERMINATION, HOPE, SUCCESS, and on SUBCONSCIENT WORK. There is an incidental remark on animal psychology; see WORKING BACKWARDS.

It is emphasized that all sorts of problems, especially PRACTICAL PROBLEMS, and even PUZZLES, are within the scope of heuristic. It is also emphasized that infallible RULES OF DISCOVERY are beyond the scope of serious research. Heuristic discusses human behavior in the face of problems; this has been in fashion, presumably, since the beginning of human society, and the quintessence of such ancient discussions seems to be preserved in the WISDOM OF PROVERBS.

4. A few articles on particular questions are included and some articles on more general aspects are expanded, because they could be, or parts of them could be, of special interest to students or teachers.

There are articles discussing methodical questions often important in elementary mathematics, as PAPPUS, WORKING BACKWARDS (already quoted under 3), REDUCTIO AD ABSURDUM AND INDIRECT PROOF, INDUCTION AND MATHEMATICAL INDUCTION, SETTING UP EQUATIONS, TEST BY DIMENSION, and WHY PROOFS? A few articles address themselves more particularly to teachers, as ROUTINE PROBLEMS and DIAGNOSIS, and others to students somewhat more ambitious than the average, as THE INTELLIGENT PROBLEM-SOLVER, THE INTELLIGENT READER, and THE FUTURE MATHEMATICIAN.

It may be mentioned here that the dialogues between the teacher and his students, given in sections **8, 10, 18, 19, 20** and in various articles of the Dictionary may serve as models not only to the teacher who tries to guide his class but also to the problem-solver who works by himself. To describe thinking as "mental discourse," as a sort of conversation of the thinker with himself, is not inappropriate. The dialogues in question show the progress of the solution; the problem-solver, talking with himself, may progress along a similar line.

5. We are not going to exhaust the remaining titles; just a few groups will be mentioned.

Some articles contain remarks on the history of our subject, on DESCARTES, LEIBNITZ, BOLZANO, on HEURISTIC, on TERMS, OLD AND NEW and on PAPPUS (this last one has been quoted already under 4).

A few articles explain technical terms: CONDITION, COROLLARY, LEMMA.

Some articles contain only cross-references (they are marked with daggers [†] in the Table of Contents).

6. Heuristic aims at generality, at the study of procedures which are independent of the subject-matter and apply to all sorts of problems. The present exposition, however, quotes almost exclusively elementary mathe-

matical problems as examples. It should not be overlooked that this is a restriction but it is hoped that this restriction does not impair seriously the trend of our study. In fact, elementary mathematical problems present all the desirable variety, and the study of their solution is particularly accessible and interesting. Moreover, nonmathematical problems although seldom quoted as examples are never completely forgotten. More advanced mathematical problems are never directly quoted but constitute the real background of the present exposition. The expert mathematician who has some interest for this sort of study can easily add examples from his own experience to elucidate the points illustrated by elementary examples here.

7. The writer of this book wishes to acknowledge his indebtedness and express his gratitude to a few modern authors, not quoted in the article on HEURISTIC. They are the physicist and philosopher Ernst Mach, the mathematician Jacques Hadamard, the psychologists William James and Wolfgang Köhler. He wishes also to quote the psychologist K. Duncker and the mathematician F. Krauss whose work (published after his own research was fairly advanced, and partly published) shows certain parallel remarks.

Notation. If you wish to realize the advantages of a well chosen and well known notation try to add a few not too small numbers with the condition that you are not allowed to use the familiar Arabic numerals, although you may use, if you wish to write, Roman numerals. Take, for instance, the numbers MMMXC, MDXCVI, MDCXLVI, MDCCLXXXI, MDCCCLXXXVII.

We can scarcely overestimate the importance of mathematical notation. Modern computers, using the decimal notation, have a great advantage over the ancient computers who did not have such a convenient manner of writing the numbers. An average modern student who is familiar with the usual notation of algebra, analytical geometry, and the differential and integral calculus, has an immense advantage over a Greek mathematician in solving the problems about areas and volumes which exercised the genius of Archimedes.

1. Speaking and thinking are closely connected, the use of words assists the mind. Certain philosophers and philologists went a little further and asserted that the use of words is indispensable to the use of reason.

Yet this last assertion appears somewhat exaggerated. If we have a little experience of serious mathematical work we know that we can do a piece of pretty hard thinking without using any words, just looking at geometric figures or manipulating algebraic symbols. Figures and symbols are closely connected with mathematical thinking, their use assists the mind. We could improve that somewhat narrow assertion of philosophers and philologists by bringing the words into line with other sorts of signs and saying that the *use of signs appears to be indispensable to the use of reason.*

At any rate, the use of mathematical symbols is similar to the use of words. Mathematical notation appears as a sort of language, *une langue bien faite,* a language well adapted to its purpose, concise and precise, with rules which, unlike the rules of ordinary grammar, suffer no exception.

If we accept this viewpoint, SETTING UP EQUATIONS appears as a sort of translation, translation from ordinary language into the language of mathematical symbols.

2. Some mathematical symbols, as $+$, $-$, $=$, and several others, have a fixed traditional meaning, but other symbols, as the small and capital letters of the Roman and Greek alphabets, are used in different meanings in dif-

ferent problems. When we face a new problem, we must choose certain symbols, we have to *introduce suitable notation*. There is something analogous in the use of ordinary language. Many words are used in different meanings in different contexts; when precision is important, we have to choose our words carefully.

An important step in solving a problem is to choose the notation. It should be done carefully. The time we spend now on choosing the notation may be well repaid by the time we save later by avoiding hesitation and confusion. Moreover, choosing the notation carefully, we have to think sharply of the elements of the problem which must be denoted. Thus, choosing a suitable notation may contribute essentially to understanding the problem.

3. A good notation should be unambiguous, pregnant, easy to remember; it should avoid harmful second meanings, and take advantage of useful second meanings; the order and connection of signs should suggest the order and connection of things.

4. Signs must be, first of all, *unambiguous*. It is inadmissible that the same symbol denote two different objects in the same inquiry. If, solving a problem, you call a certain magnitude a you should avoid calling anything else a which is connected with the same problem. Of course, you may use the letter a in a different meaning in a different problem.

Although it is forbidden to use the same symbol for different objects it is not forbidden to use different symbols for the same object. Thus, the product of a and b may be written as

$$a \times b \qquad a \cdot b \qquad ab.$$

In some cases, it is advantageous to use two or more different signs for the same object, but such cases require particular care. Usually, it is better to use just one sign for one object, and in no case should several signs be used wantonly.

5. A good sign should be *easy to remember* and easy to recognize; the sign should immediately remind us of the object and the object of the sign.

A simple device to make signs easily recognizable is to use *initials* as symbols. For example, in section **20** we used r for rate, t for time, V for volume. We cannot use, however, initials in all cases. Thus, in section **20**, we had to consider a radius but we could not call it r because this letter was already taken to denote a rate. There are still other motives restricting the choice of symbols, and other means to make them easily recognizable which we are going to discuss.

6. Notation is not only easily recognizable but particularly helpful in shaping our conception when *the order and connection of the signs suggest the order and connection of the objects.* We need several examples to illustrate this point.

(I) In order to denote objects which are near to each other in the conception of the problem we use letters which are near to each other in the alphabet.

Thus, we generally use letters at the beginning of the alphabet as a, b, c, for given quantities or constants, and letters at the end of the alphabet as x, y, z, for unknown quantities or variables.

In section **8** we used a, b, c for the given length, width, and height of a parallelepiped. On this occasion, the notation a, b, c was preferable to the notation by initials l, w, h. The three lengths played the same role in the problem which is emphasized by the use of successive letters. Moreover, being at the beginning of the alphabet, a, b, c are, as we just said, the most usual letters to denote given quantities. On some other occasion, if the three lengths play different roles and it is important to know which lengths are horizontal and which one is vertical, the notation l, w, h might be preferable.

(II) In order to denote objects belonging to the same category, we frequently choose letters belonging to the same alphabet for one category, using different alphabets for different categories. Thus, in plane geometry we often use:

Roman capitals as $A, B, C, \ldots$ for points,
small Roman letters as $a, b, c, \ldots$ for lines,
small Greek letters as $\alpha, \beta, \gamma, \ldots$ for angles.

If there are two objects belonging to different categories but having some particular relation to each other which is important for our problem, we may choose, to denote these two objects, corresponding letters of the respective alphabets as A and a, B and b, and so on. A familiar example is the usual notation for a triangle:

A, B, C stand for the vertices,
a, b, c for the sides,
α, β, γ for the angles.

It is understood that a is the side opposite to the vertex A and the angle at A is called α.

(III) In section 20, the letters a, b, x, y are particularly well chosen to indicate the nature and connection of the elements denoted. The letters a, b hint that the magnitudes denoted are constants; x, y indicate variables; a precedes b as x precedes y and this suggests that a is in the same relation to b as x is to y. In fact, a and x are horizontal, b and y vertical, and $a : b = x : y$.

7. The notation

$$\triangle ABC \sim \triangle EFG$$

indicates that the two triangles in question are similar. In modern books, the formula is meant to indicate that the two triangles are similar, the vertices corresponding to each other in the order as they are written, A to E, B to F, C to G. In older books, this proviso about the order was not yet introduced; the reader had to look at the figure or remember the derivation in order to ascertain which vertex corresponded to which.

The modern notation is much preferable to the older one. Using the modern notation, we may draw consequences from the formula *without looking at the figure.* Thus, we may derive that

$$\angle A = \angle E$$
$$AB : BC = EF : FG$$

and other relations of the same kind. The older notation expresses less and does not allow such definite consequences.

A notation expressing more than another may be termed more *pregnant.* The modern notation for similitude of triangles is more pregnant than the older one, reflects the order and connection of things more fully than the older one, and therefore, it may serve as basis for more consequences than the older one.

8. Words have *second meanings.* Certain contexts in which a word is often used influence it and add something to its primary meaning, some shade, or second meaning, or "connotation." If we write carefully, we try to choose among the words having almost the same meaning the one whose second meaning is best adapted.

There is something similar in mathematical notation. Even mathematical symbols may acquire a sort of second meaning from contexts in which they are often used. If we choose our notation carefully, we have to take this circumstance into account. Let us illustrate the point.

There are certain letters which have acquired a firmly rooted, traditional meaning. Thus, e stands usually for the basis of natural logarithms, i for $\sqrt{-1}$, the imaginary unit, and π for the ratio of the circumference of the circle to the diameter. It is on the whole better to use such symbols only in their traditional meaning. If we use such a symbol in some other meaning its traditional meaning could occasionally interfere and be embarrassing, even misleading. It is true that harmful second meanings of this sort give less trouble to the beginner who has not yet studied many subjects than to the mathematician who should have sufficient experience to deal with such nuisances.

Second meanings of the symbols can also be helpful, even very helpful, if they are used with tact. A notation used on former occasions may assist us in recalling some useful procedure; of course, we should be sufficiently careful to separate clearly the present (primary) meaning of the symbol from its former (secondary) meaning. A *standing notation* [as the traditional notation for the parts of the triangle which we mentioned before, 6 (II)] has great advantages; used on several former occasions it may assist us in recalling various formerly used procedures. We remember our formulas in some standing notation. Of course, we should be sufficiently careful when, owing to particular circumstances, we are obliged to use a standing notation in a meaning somewhat different from the usual one.

9. When we have to choose between two notations, one reason may speak for one, and some other reason for the other. We need experience and taste to choose the more suitable notation as we need experience and taste to choose more suitable words. Yet it is good to know the various advantages and disadvantages discussed in the foregoing. At any rate, we should choose our notation carefully, and have *some good reason* for our choice.

10. Not only the most hopeless boys in the class but

also quite intelligent students may have an aversion for algebra. There is always something arbitrary and artificial about notation; to learn a new notation is a burden for the memory. The intelligent student refuses to assume the burden if he does not see any compensation for it. The intelligent student is justified in his aversion for algebra if he is not given ample opportunity to convince himself by his own experience that *the language of mathematical symbols assists the mind.* To help him to such experience is an important task of the teacher, one of his most important tasks.

I say that it is an important task but I do not say that it is an easy one. The foregoing remarks may be of some help. See also SETTING UP EQUATIONS. Checking a formula by extensive discussion of its properties may be recommended as a particularly instructive exercise; see section 14 and CAN YOU CHECK THE RESULT? 2.

Pappus, an important Greek mathematician, lived probably around A.D. 300. In the seventh book of his *Collectiones,* Pappus reports about a branch of study which he calls *analyomenos.* We can render this name in English as "Treasury of Analysis," or as "Art of Solving Problems," or even as "Heuristic"; the last term seems to be preferable here. A good English translation of Pappus's report is easily accessible[7]; what follows is a free rendering of the original text:

"The so-called Heuristic is, to put it shortly, a special body of doctrine for the use of those who, after having studied the ordinary Elements, are desirous of acquiring the ability to solve mathematical problems, and it is useful for this alone. It is the work of three men, Euclid, the author of the Elements, Apollonius of Perga, and

7 T. L. Heath, *The Thirteen Books of Euclid's Elements,* Cambridge, 1908, vol. 1, p. 138.

Aristaeus the elder. It teaches the procedures of analysis and synthesis.

"In analysis, we start from what is required, we take it for granted, and we draw consequences from it, and consequences from the consequences, till we reach a point that we can use as starting point in synthesis. For in analysis we assume what is required to be done as already done (what is sought as already found, what we have to prove as true). We inquire from what antecedent the desired result could be derived; then we inquire again what could be the antecedent of that antecedent, and so on, until passing from antecedent to antecedent, we come eventually upon something already known or admittedly true. This procedure we call analysis, or solution backwards, or regressive reasoning.

"But in synthesis, reversing the process, we start from the point which we reached last of all in the analysis, from the thing already known or admittedly true. We derive from it what preceded it in the analysis, and go on making derivations until, retracing our steps, we finally succeed in arriving at what is required. This procedure we call synthesis, or constructive solution, or progressive reasoning.

"Now analysis is of two kinds; the one is the analysis of the 'problems to prove' and aims at establishing true theorems; the other is the analysis of the 'problems to find' and aims at finding the unknown.

"If we have a 'problem to prove' we are required to prove or disprove a clearly stated theorem A. We do not know yet whether A is true or false; but we derive from A another theorem B, from B another C, and so on, until we come upon a last theorem L about which we have definite knowledge. If L is true, A will be also true, provided that all our derivations are convertible. From L we prove the theorem K which preceded L in the analysis and, proceding in the same way, we retrace our steps; from C we prove B, from B we prove A, and so we attain our aim. If, however, L is false, we have proved A false.

"If we have a 'problem to find' we are required to find a certain unknown x satisfying a clearly stated condition. We do not know yet whether a thing satisfying such a condition is possible or not; but assuming that there is an x satisfying the condition imposed we derive from it another unknown y which has to satisfy a related condition; then we link y to still another unknown, and so on, until we come upon a last unknown z which we can find by some known method. If there is actually a z satisfying the condition imposed upon it, there will be also an x satisfying the original condition, provided that all our derivations are convertible. We first find z; then, knowing z, we find the unknown that preceded z in the analysis; proceeding in the same way, we retrace our steps, and finally, knowing y, we obtain x, and so we attain our aim. If, however, there is nothing that would satisfy the condition imposed upon z, the problem concerning x has no solution."

We should not forget that the foregoing is not a literal translation but a free rendering, a *paraphrase*. Various differences between the original and the paraphrase deserve comment, for Pappus's text is important in many ways.

1. Our paraphrase uses a more definite terminology than the original and introduces the symbols $A, B, \ldots L, x, y, \ldots z$ which the original has not.

2. The paraphrase has (p. 141, line 30) "mathematical problems" where the original means "geometrical problems." This emphasizes that the procedures described by Pappus are by no means restricted to geometric problems; they are, in fact, not even restricted to mathematical problems. We have to illustrate this by examples

since, in these matters, generality and independence from the nature of the subject are important (see section 3).

3. *Algebraic illustration.* Find x satisfying the equation

$$8(4^x + 4^{-x}) - 54(2^x + 2^{-x}) + 101 = 0.$$

This is a "problem to find," not too easy for a beginner. He has to be familiar with the idea of analysis; not with the word "analysis" of course, but with the idea of attaining the aim by repeated reduction. Moreover, he has to be familiar with the simplest sorts of equations. Even with some knowledge, it takes a good idea, a little luck, a little invention to observe that, since $4^x = (2^x)^2$ and $4^{-x} = (2^x)^{-2}$, it may be advantageous to introduce

$$y = 2^x.$$

Now, this substitution is really advantageous, the equation obtained for y

$$8\left(y^2 + \frac{1}{y^2}\right) - 54\left(y + \frac{1}{y}\right) + 101 = 0$$

appears simpler than the original equation; but our task is not yet finished. It needs another little invention, another substitution

$$z = y + \frac{1}{y}$$

which transforms the condition into

$$8z^2 - 54z + 85 = 0.$$

Here the analysis ends, provided that the problem-solver is acquainted with the solution of quadratic equations.

What is the synthesis? Carrying through, step by step, the calculations whose possibility was foreseen by the analysis. The problem-solver needs no new idea to finish his problem, only some patience and attention in calculating the various unknowns. The order of calculation is opposite to the order of invention; first z is found ($z = 5/2, 17/4$), then y ($y = 2, 1/2, 4, 1/4$), and finally the originally required x ($x = 1, -1, 2, -2$). The synthesis retraces the steps of the analysis, and it is easy to see in the present case why it does so.

4. *Nonmathematical illustration.* A primitive man wishes to cross a creek; but he cannot do so in the usual way because the water has risen overnight. Thus, the crossing becomes the object of a problem; "crossing the creek" is the x of this primitive problem. The man may recall that he has crossed some other creek by walking along a fallen tree. He looks around for a suitable fallen tree which becomes his new unknown, his y. He cannot find any suitable tree but there are plenty of trees standing along the creek; he wishes that one of them would fall. Could he make a tree fall across the creek? There is a great idea and there is a new unknown; by what means could he tilt the tree over the creek?

This train of ideas ought to be called analysis if we accept the terminology of Pappus. If the primitive man succeeds in finishing his analysis he may become the inventor of the bridge and of the axe. What will be the synthesis? Translation of ideas into actions. The finishing act of the synthesis is walking along a tree across the creek.

The same objects fill the analysis and the synthesis; they exercise the mind of the man in the analysis and his muscles in the synthesis; the analysis consists in thoughts, the synthesis in acts. There is another difference; the order is reversed. Walking across the creek is the first desire from which the analysis starts and it is the last act with which the synthesis ends.

5. The paraphrase hints a little more distinctly than the original the natural connection between analysis and synthesis. This connection is manifest after the foregoing

examples. Analysis comes naturally first, synthesis afterwards; analysis is invention, synthesis, execution; *analysis is devising a plan, synthesis carrying through the plan.*

6. The paraphrase preserves and even emphasizes certain curious phrases of the original: "assume what is required to be done as already done, what is sought as found, what you have to prove as true." This is paradoxical; is it not mere self-deception to assume that the problem that we have to solve is solved? This is obscure; what does it mean? If we consider closely the context and try honestly to understand our own experience in solving problems, the meaning can scarcely be doubtful.

Let us first consider a "problem to find." Let us call the unknown x and the data a, b, c. To "assume the problem as solved" means to assume that there exists an object x satisfying the condition—that is, having those relations to the data a, b, c which the condition prescribes. This assumption is made just in order to start the analysis, it is provisional, and it is harmless. For, *if* there is no such object *and* the analysis leads us anywhere, it is bound to lead us to a final problem that has no solution and hence it will be manifest that our original problem has no solution. Then, the assumption is useful. In order to examine the condition, we have to conceive, to represent to ourselves, or to visualize geometrically the relations which the condition prescribes between x and a, b, c; how could we do so without conceiving, representing, or visualizing x as existent? Finally, the assumption is natural. The primitive man whose thoughts and deeds we discussed in comment 4 imagines himself walking on a fallen tree and crossing the creek long before he actually can do so; he sees his problem "as solved."

The object of a "problem to prove" is to prove a certain theorem A. The advice to "assume A as true" is just an invitation to draw consequences from the theorem A

although we have not yet proved it. People with a certain mental character or a certain philosophy may shrink from drawing consequences from an unproved theorem; but such people cannot start an analysis.

Compare FIGURES, 2.

7. The paraphrase uses twice the important phrase "provided that all our derivations are convertible"; see p. 142, line 33 and p. 143, lines 14–15. This is an interpolation; the original contains nothing of the sort and the lack of such a proviso was observed and criticized in modern times. See AUXILIARY PROBLEM, 6 for the notion of "convertible reduction."

8. The "analysis of the problems to prove" is explained in the paraphrase in words quite different from those used by the original but there is no change in the sense; at any rate, there is no intention to change the sense. The analysis of the "problem to find," however, is explained more concretely in the paraphrase than in the original. The original seems to aim at the description of a somewhat more general procedure, the construction of a *chain of equivalent auxiliary problems* which is described in AUXILIARY PROBLEM, 7.

9. Many elementary textbooks of geometry contain a few remarks about analysis, synthesis, and "assuming the problem as solved." There is little doubt that this almost ineradicable tradition goes back to Pappus, although there is hardly a current textbook whose writer would show any direct acquaintance with Pappus. The subject is important enough to be mentioned in elementary textbooks but easily misunderstood. The circumstance alone that it is restricted to textbooks of geometry shows a current lack of understanding; see comment 2 above. If the foregoing comments could contribute to a better understanding of this matter their length would be amply justified.

For another example, a different viewpoint, and further comments see WORKING BACKWARDS.

Compare also REDUCTIO AD ABSURDUM AND INDIRECT PROOF, 2.

Pedantry and mastery are opposite attitudes toward rules.

1. To apply a rule to the letter, rigidly, unquestioningly, in cases where it fits and in cases where it does not fit, is pedantry. Some pedants are poor fools; they never did understand the rule which they apply so conscientiously and so indiscriminately. Some pedants are quite successful; they understood their rule, at least in the beginning (before they became pedants), and chose a good one that fits in many cases and fails only occasionally.

To apply a rule with natural ease, with judgment, noticing the cases where it fits, and without ever letting the words of the rule obscure the purpose of the action or the opportunities of the situation, is mastery.

2. The questions and suggestions of our list may be helpful both to problem-solvers and to teachers. But, first, they must be understood, their proper use must be learned, and learned by trial and error, by failure and success, by experience in applying them. Second, their use should never become pedantic. You should ask no question, make no suggestion, indiscriminately, following some rigid habit. Be prepared for various questions and suggestions and use your judgment. You are doing a hard and exciting problem; the step you are going to try next should be prompted by an attentive and open-minded consideration of the problem before you. You wish to help a student; what you say to your student should proceed from a sympathetic understanding of his difficulties.

And if you are inclined to be a pedant and must rely upon some rule learn this one: Always use your own brains first.

Practical problems are different in various respects from purely mathematical problems, yet the principal motives and procedures of the solution are essentially the same. Practical engineering problems usually involve mathematical problems. We will say a few words about the differences, analogies, and connections between these two sorts of problems.

1. An impressive practical problem is the construction of a dam across a river. We need no special knowledge to understand this problem. In almost prehistoric times, long before our modern age of scientific theories, men built dams of some sort in the valley of the Nile, and in other parts of the world, where the crops depended on irrigation.

Let us visualize the problem of constructing an important modern dam.

What is the unknown? Many unknowns are involved in a problem of this kind: the exact location of the dam, its geometric shape and dimensions, the materials used in its construction, and so on.

What is the condition? We cannot answer this question in one short sentence because there are many conditions. In so large a project it is necessary to satisfy many important economic needs and to hurt other needs as little as possible. The dam should provide electric power, supply water for irrigation or the use of certain communities, and also help to control floods. On the other hand, it should disturb as little as possible navigation, or economically important fish-life, or beautiful scenery; and so forth. And, of course, it should cost as little as possible and be constructed as quickly as possible.

What are the data? The multitude of desirable data is tremendous. We need topographical data concerning the vicinity of the river and its tributaries; geological data important for the solidity of foundations, possible leakage, and available materials of construction; meteorological data about annual precipitation and the height of floods; economic data concerning the value of ground which will be flooded, cost of materials and labor; and so on.

Our example shows that unknowns, data, and conditions are more complex and less sharply defined in a practical problem than in a mathematical problem.

2. In order to solve a problem, we need a certain amount of previously acquired knowledge. The modern engineer has a highly specialized body of knowledge at his disposal, a scientific theory of the strength of materials, his own experience, and the mass of engineering experience stored in special technical literature. We cannot avail ourselves of such special knowledge here but we may try to imagine what was in the mind of an ancient Egyptian dam-builder.

He has seen, of course, various other, perhaps smaller, dams: banks of earth or masonry holding back the water. He has seen the flood, laden with all sorts of debris, pressing against the bank. He might have helped to repair the cracks and the erosion left by the flood. He might have seen a dam break, giving way under the impact of the flood. He has certainly heard stories about dams withstanding the test of centuries or causing catastrophe by an unexpected break. His mind may have pictured the pressure of the river against the surface of the dam and the strain and stress in its interior.

Yet the Egyptian dam-builder had no precise, quantitative, scientific concepts of fluid pressure or of strain and stress in a solid body. Such concepts form an essential part of the intellectual equipment of a modern engineer. Yet the latter also uses much knowledge which has not yet quite reached a precise, scientific level; what he knows about erosion by flowing water, the transportation of silt, the plasticity and other not quite clearly circumscribed properties of certain materials, is knowledge of a rather empirical character.

Our example shows that the knowledge needed and the concepts used are more complex and less sharply defined in practical problems than in mathematical problems.

3. Unknowns, data, conditions, concepts, necessary preliminary knowledge, everything is more complex and less sharp in practical problems than in purely mathematical problems. This is an important difference, perhaps the main difference, and it certainly implies further differences; yet the fundamental motives and procedures of the solution appear to be the same for both sorts of problems.

There is a widespread opinion that practical problems need more experience than mathematical problems. This may be so. Yet, very likely, the difference lies in the nature of the knowledge needed and not in our attitude toward the problem. In solving a problem of one or the other kind, we have to rely on our experience with similar problems and we often ask the questions: *Have you seen the same problem in a slightly different form? Do you know a related problem?*

In solving a mathematical problem, we start from very clear concepts which are fairly well ordered in our mind. In solving a practical problem, we are often obliged to start from rather hazy ideas; then, the clarification of the concepts may become an important part of the problem. Thus, medical science is in a better position to check infectious diseases today than it was in the times before Pasteur when the notion of infection itself was rather hazy. *Have you taken into account all essential notions involved in the problem?* This is a good question for all sorts of problems but its use varies widely with the nature of the intervening notions.

In a perfectly stated mathematical problem all data and all clauses of the condition are essential and must be taken into account. In practical problems we have a multitude of data and conditions; we take into account as many as we can but we are obliged to neglect some. Take the case of the designer of a large dam. He considers the public interest and important economic interests but he is bound to disregard certain petty claims and grievances. The data of his problem are, strictly speaking, inexhaustible. For instance, he would like to know a little more about the geologic nature of the ground on which the foundations must be laid, but eventually he must stop collecting geologic data although a certain margin of uncertainty unavoidably remains.

Did you use all the data? Did you use the whole condition? We cannot miss these questions when we deal with purely mathematical problems. In practical problems, however, we should put these questions in a modified form: Did you use all the data which *could contribute appreciably* to the solution? Did you use all the conditions which *could influence appreciably* the solution? We take stock of the available relevant information, we collect more information if necessary, but eventually we must stop collecting, we must draw the line somewhere, we cannot help neglecting something. "If you will sail without danger, you must never put to sea." Quite often, there is a great surplus of data which have no appreciable influence on the final form of the solution.

4. The designers of the ancient Egyptian dams had to rely on the common-sense interpretation of their experi-

ence, they had nothing else to rely on. The modern engineer cannot rely on common sense alone, especially if his project is of a new and daring design; he has to calculate the resistance of the projected dam, foresee quantitatively the strain and stress in its interior. For this purpose, he has to apply the theory of elasticity (which applies fairly well to constructions in concrete). To apply this theory, he needs a good deal of mathematics; the practical engineering problem leads to a mathematical problem.

This mathematical problem is too technical to be discussed here; all we can say about it is a general remark. In setting up and in solving mathematical problems derived from practical problems, we usually content ourselves with an *approximation*. We are bound to neglect some minor data and conditions of the practical problem. Therefore it is reasonable to allow some slight inaccuracy in the computations especially when we can gain in simplicity what we lose in accuracy.

5. Much could be said about approximations that would deserve general interest. We cannot suppose, however, any specialized mathematical knowledge and therefore we restrict ourselves to just one intuitive and instructive example.

The drawing of geographic maps is an important practical problem. Devising a map, we often assume that the earth is a sphere. Now this is only an approximate assumption and not the exact truth. The surface of the earth is not at all a mathematically defined surface and we definitely know that the earth is flattened at the poles. Assuming, however, that the earth is a sphere, we may draw a map of it much more easily. We gain much in simplicity and do not lose a great deal in accuracy. In fact, let us imagine a big ball that has exactly the shape of the earth and that has a diameter of 25 feet at its equator. The distance between the poles of such a ball is less than 25 feet because the earth is flattened, but only about one inch less. Thus the sphere yields a good practical approximation.

Problems to find, problems to prove. We draw a parallel between these two kinds of problems.

1. The aim of a "problem to find" is to find a certain object, the unknown of the problem.

The unknown is also called "quaesitum," or the thing sought, or the thing required. "Problems to find" may be theoretical or practical, abstract or concrete, serious problems or mere puzzles. We may seek all sorts of unknowns; we may try to find, to obtain, to acquire, to produce, or to construct all imaginable kinds of objects. In the problem of the mystery story the unknown is a murderer. In a chess problem the unknown is a move of the chessmen. In certain riddles the unknown is a word. In certain elementary problems of algebra the unknown is a number. In a problem of geometric construction the unknown is a figure.

2. The aim of a "problem to prove" is to show conclusively that a certain clearly stated assertion is true, or else to show that it is false. We have to answer the question: Is this assertion true or false? And we have to answer conclusively, either by proving the assertion true, or by proving it false.

A witness affirms that the defendant stayed at home a certain night. The judge has to find out whether this assertion is true or not and, moreover, he has to give as good grounds as possible for his finding. Thus, the judge has a "problem to prove." Another "problem to prove" is to "prove the theorem of Pythagoras." We do not say: "Prove or disprove the theorem of Pythagoras." It would be better in some respects to include in the statement of the problem the possibility of disproving, but we may neglect it, because we know that the chances for disproving the theorem of Pythagoras are rather slight.

3. The principal parts of a "problem to find" are the *unknown,* the *data,* and the *condition.*

If we have to construct a triangle with sides $a, b, c,$ the unknown is a triangle, the data are the three lengths $a, b, c,$ and the triangle is required to satisfy the condition that its sides have the given lengths $a, b, c.$ If we have to construct a triangle whose altitudes are $a, b, c,$ the unknown is an object of the same category as before, the data are the same, but the condition linking the unknown to the data is different.

4. If a "problem to prove" is a mathematical problem of the usual kind, its principal parts are the *hypothesis* and the *conclusion* of the theorem which has to be proved or disproved.

"If the four sides of a quadrilateral are equal, then the two diagonals are perpendicular to each other." The second part starting with "then" is the conclusion, the first part starting with "if" is the hypothesis.

[Not all mathematical theorems can be split naturally into hypothesis and conclusion. Thus, it is scarcely possible to split so the theorem: "There are an infinity of prime numbers."]

5. If you wish to solve a "problem to find" you must know, and know very exactly, its principal parts, the unknown, the data, and the condition. Our list contains many questions and suggestions concerned with these parts.

What is the unknown? What are the data? What is the condition?

Separate the various parts of the condition.

Find the connection between the data and the unknown.

Look at the unknown! And try to think of a familiar problem having the same or a similar unknown.

Keep only a part of the condition, drop the other part; how far is the unknown then determined, how can it vary? Could you derive something useful from the data? Could you think of other data appropriate to determine the unknown? Could you change the unknown, or the data, or both if necessary, so that the new unknown and the new data are nearer to each other?

Did you use all the data? Did you use the whole condition?

6. If you wish to solve a "problem to prove" you must know, and know very exactly, its principal parts, the hypothesis, and the conclusion. There are useful questions and suggestions concerning these parts which correspond to those questions and suggestions of our list which are specially adapted to "problems to find."

What is the hypothesis? What is the conclusion?

Separate the various parts of the hypothesis.

Find the connection between the hypothesis and the conclusion.

Look at the conclusion! And try to think of a familiar theorem having the same or a similar conclusion.

Keep only a part of the hypothesis, drop the other part; is the conclusion still valid? Could you derive something useful from the hypothesis? Could you think of another hypothesis from which you could easily derive the conclusion? Could you change the hypothesis, or the conclusion, or both if necessary, so that the new hypothesis and the new conclusion are nearer to each other?

Did you use the whole hypothesis?

7. "Problems to find" are more important in elementary mathematics, "problems to prove" more important in advanced mathematics. In the present book, "problems to find" are more emphasized than the other kind,

but the author hopes to reestablish the balance in a fuller treatment of the subject.

Progress and achievement. Have you made any progress? What was the essential achievement? We may address questions of this kind to ourselves when we are solving a problem or to a student whose work we supervise. Thus, we are used to judge, more or less confidently, progress and achievement in concrete cases. The step from such concrete cases to a general description is not easy at all. Yet we have to undertake this step if we wish to make our study of heuristic somewhat complete and we must try to clarify what constitutes, in general, progress and achievement in solving problems.

1. In order to solve a problem, we must have some knowledge of the subject-matter and we must select and collect the relevant items of our existing but initially dormant knowledge. There is much more in our conception of the problem at the end than was in it at the outset; what has been added? What we have succeeded in extracting from our memory. In order to obtain the solution we have to recall various essential facts. We have to recollect formerly solved problems, known theorems, definitions, if our problem is mathematical. Extracting such relevant elements from our memory may be termed *mobilization.*

2. In order to solve a problem, however, it is not enough to recollect isolated facts, we must combine these facts, and their combination must be well adapted to the problem at hand. Thus, in solving a mathematical problem, we have to construct an argument connecting the materials recollected to a well adapted whole. This adapting and combining activity may be termed *organization.*

3. In fact, mobilization and organization can never be

really separated. Working at the problem with concentration, we recall only facts which are more or less connected with our purpose, and we have nothing to connect and organize but materials we have recollected and mobilized.

Mobilization and organization are but two *aspects* of the same complex process which has still many other aspects.

4. Another aspect of the progress of our work is that our *mode of conception changes.* Enriched with all the materials which we have recalled, adapted to it, and worked into it, our conception of the problem is much fuller at the end than it was at the outset. Desiring to proceed from our initial conception of the problem to a more adequate, better adapted conception, we try various standpoints and view the problem from different sides. We could make hardly any progress without VARIATION OF THE PROBLEM.

5. As we progress toward our final goal we see more and more of it, and when we see it better we judge that we are nearer to it. As our examination of the problem advances, we *foresee* more and more clearly what should be done for the solution and how it should be done. Solving a mathematical problem we may foresee, if we are lucky, that a certain known theorem might be used, that the consideration of a certain formerly solved problem might be helpful, that going back to the meaning of a certain technical term might be necessary. We do not foresee such things with certainty, only with a certain degree of plausibility. We shall attain complete certainty when we have obtained the complete solution, but before obtaining certainty we must often be satisfied with a more or less plausible guess. Without considerations which are only plausible and provisional, we could never find the solution which is certain and final. We need HEURISTIC REASONING.

6. What is progress toward the solution? Advancing mobilization and organization of our knowledge, evolution of our conception of the problem, increasing prevision of the steps which will constitute the final argument. We may advance steadily, by small imperceptible steps, but now and then we advance abruptly, by leaps and bounds. A sudden advance toward the solution is called a BRIGHT IDEA, a good idea, a happy thought, a brain-wave (in German there is a more technical term, *Einfall*). What is a bright idea? An abrupt and momentous change of our outlook, a sudden reorganization of our mode of conceiving the problem, a just emerging confident prevision of the steps we have to take in order to attain the solution.

7. The foregoing considerations provide the questions and suggestions of our list with the right sort of background.

Many of these questions and suggestions aim directly at the *mobilization* of our formerly acquired knowledge: *Have you seen it before? Or have you seen the same problem in a slightly different form? Do you know a related problem? Do you know a theorem that could be useful? Look at the unknown! And try to think of a familiar problem having the same or a similar unknown.*

There are typical situations in which we think that we have collected the right sort of material and we work for a better *organization* of what we have mobilized: *Here is a problem related to yours and solved before. Could you use it? Could you use its result? Could you use its method? Should you introduce some auxiliary element in order to make its use possible?*

There are other typical situations in which we think that we have not yet collected enough material. We wonder what is missing: *Did you use all the data? Did you use the whole condition? Have you taken into account all essential notions involved in the problem?*

Some questions aim directly at the *variation* of the problem: *Could you restate the problem? Could you restate it still differently?* Many questions aim at the variation of the problem by specified means, as going back to the DEFINITION, using ANALOGY, GENERALIZATION, SPECIALIZATION, DECOMPOSING AND RECOMBINING.

Still other questions suggest a trial to *foresee* the nature of the solution we are striving to obtain: *Is it possible to satisfy the condition? Is the condition sufficient to determine the unknown? Or is it insufficient? Or redundant? Or contradictory?*

The questions and suggestions of our list do not mention directly the *bright idea*; but, in fact, all are concerned with it. Understanding the problem we prepare for it, devising a plan we try to provoke it, having provoked it we carry it through, looking back at the course and the result of the solution we try to exploit it better.[8]

Puzzles. According to section 3, the questions and suggestions of our list are independent of the subject-matter and applicable to all kinds of problems. It is quite interesting to test this assertion on various puzzles.

Take, for instance, the words

DRY OXTAIL IN REAR.

The problem is to find an "anagram," that is, a rearrangement of the letters contained in the given words into one word. It is interesting to observe that, when we are solving this puzzle, several questions of our list are pertinent and even stimulating.

What is the unknown? A word.

What are the data? The four words DRY OXTAIL IN REAR.

[8] Several points discussed in this article are more fully considered in the author's paper, *Acta Psychologica,* vol. 4 (1938), pp. 113-170.

What is the condition? The desired word has fifteen letters, the letters contained in the four given words. It is probably a not too unusual English word.

Draw a figure. It is quite useful to mark out fifteen blank spaces:

- - - - - - - - - - - - - - -

Could you restate the problem? We have to find a word containing, in some arrangement, the letters

AAEIIOY DLNRRRTX.

This is certainly an equivalent restatement of the problem (see AUXILIARY PROBLEM, 6). It may be an advantageous restatement. Separating the vowels from the consonants (this is important, the alphabetical order is not) we see another aspect of the problem. Thus, we see now that the desired word has seven syllables unless it has some diphthongs.

If you cannot solve the proposed problem, try to solve first some related problem. A related problem is to form words with some of the given letters. We can certainly form short words of this kind. Then we try to find longer and longer words. The more letters we use the nearer we may come to the desired word.

Could you solve a part of the problem? The desired word is so long that it must have distinct parts. It is, probably, a compound word, or it is derived from some other word by adding some usual ending. Which usual ending could it be?

- - - - - - - - - - A T I O N
- - - - - - - - - - - - E L Y

Keep only a part of the condition and drop the other part. We may try to think of a long word with, possibly, as many as seven syllables and relatively few consonants, containing an X and a Y.

The questions and suggestions of our list cannot work magic. They cannot give us the solution of all possible puzzles without any effort on our part. If the reader wishes to find the word, he must keep on trying and thinking about it. What the questions and suggestions of the list can do is to "keep the ball rolling." When, discouraged by lack of success, we are inclined to drop the problem, they may suggest to us a new trial, a new aspect, a new variation of the problem, a new stimulus; they may keep us thinking.

For another example see DECOMPOSING AND RECOMBINING, 8.

Reductio ad absurdum and indirect proof are different but related procedures.

Reductio ad absurdum shows the falsity of an assumption by deriving from it a manifest absurdity. "Reduction to an absurdity" is a mathematical procedure but it has some resemblance to irony which is the favorite procedure of the satirist. Irony adopts, to all appearance, a certain opinion and stresses it and overstresses it till it leads to a manifest absurdity.

Indirect proof establishes the truth of an assertion by showing the falsity of the opposite assumption. Thus, indirect proof has some resemblance to a politician's trick of establishing a candidate by demolishing the reputation of his opponent.

Both "reductio ad absurdum" and indirect proof are effective tools of discovery which present themselves naturally to an intent mind. Nevertheless, they are disliked by a few philosophers and many beginners, which is understandable; satirical people and tricky politicians do not appeal to everybody. We shall first illustrate the effectiveness of both procedures by examples and discuss objections against them afterwards.

1. *Reductio ad absurdum.* Write numbers using each of the ten digits exactly once so that the sum of the numbers is exactly 100.

We may learn something by trying to solve this puzzle whose statement demands some elucidation.

What is the unknown? A set of numbers; and by numbers we mean here, of course, ordinary integers.

What is given? The number 100.

What is the condition? The condition has two parts. First, writing the desired set of numbers, we must use each of the ten digits, 0, 1, 2, 3, 4, 5, 6, 7, 8 and 9, just once. Second, the sum of all numbers in the set must be 100.

Keep only a part of the condition, drop the other part. The first part alone is easy to satisfy. Take the set 19, 28, 37, 46, 50; each figure occurs just once. But, of course, the second part of the condition is not satisfied; the sum of these numbers is 180, not 100. We could, however, do better. "Try, try again." Yes,

$$19 + 28 + 30 + 7 + 6 + 5 + 4 = 99.$$

The first part of the condition is satisfied, and the second part is almost satisfied; we have 99 instead of 100. Of course, we can easily satisfy the second part if we drop the first:

$$19 + 28 + 31 + 7 + 6 + 5 + 4 = 100.$$

The first part is not satisfied: the figure 1 occurs twice, and 0 not at all; the other figures are all right. "Try, try again."

After a few unsuccessful trials, however, we may be led to suspect that it is not possible to obtain 100 in the manner required. Eventually the problem arises: *Prove that it is impossible to satisfy both parts of the proposed condition at the same time.*

Quite good students may find that this problem is above their heads. Yet the answer is easy enough if we have the right attitude. *We have to examine the hypothetical situation in which both parts of the condition are satisfied.*

We suspect that this situation cannot actually arise and our suspicion, based on the experience of our unsuccessful trials, has some foundation. Nevertheless, let us keep an open mind and let us face the situation in which hypothetically, supposedly, allegedly both parts of the condition are satisfied. Thus, let us imagine a set of numbers whose sum is 100. They must be numbers with one or two figures. There are ten figures, and these ten figures must be all different, since each of the figures, 0, 1, 2, . . . 9 should occur just once. Thus, the sum of all ten figures must be

$$0 + 1 + 2 + 3 + 4 + 5 + 6 + 7 + 8 + 9 = 45.$$

Some of these figures denote units and others tens. It takes a little sagacity to hit upon the idea that the *sum of the figures denoting tens* may be of some importance. In fact, let t stand for this sum. Then the sum of the remaining figures, denoting units, is $45 - t$. Therefore, the sum of all numbers in the set must be

$$10t + (45 - t) = 100.$$

We have here an equation to determine t. It is of the first degree and gives

$$t = \frac{55}{9}.$$

Now, there is something that is definitely wrong. The value of t that we have found is not an integer and t should be, of course, an integer. Starting from the supposition that both parts of the proposed condition can be simultaneously satisfied, we have been led to a manifest absurdity. How can we explain this? Our original supposition must be wrong; both parts of the condition *cannot* be satisfied at the same time. And so we have attained our goal, we have succeeded in proving that the two parts of the proposed condition are incompatible.

Our reasoning is a typical "reductio ad absurdum."

2. *Remarks.* Let us look back at the foregoing reasoning and understand its general trend.

We wish to prove that it is impossible to fulfill a certain condition, that is, that the situation in which all parts of the condition are simultaneously satisfied can never arise. But, if we have proved nothing yet, we have to face the possibility that the situation could arise. Only by facing squarely the hypothetical situation and examining it closely can we hope to perceive some definitely wrong point in it. And we must lay our hand upon some definitely wrong point if we wish to show conclusively that the situation is impossible. Hence we can see that the procedure that was successful in our example is reasonable in general: We have to examine the hypothetical situation in which all parts of the condition are satisfied, *although such a situation appears extremely unlikely.*

The more experienced reader may see here another point. The main step of our procedure consisted in setting up an equation for t. Now, we could have arrived at the same equation without suspecting that something was wrong with the condition. If we wish to set up an equation, we have to express in mathematical language that all parts of the condition are satisfied, *although we do not know yet whether it is actually possible to satisfy all these parts simultaneously.*

Our procedure is "open-minded." We may hope to find the unknown satisfying the condition, or we may hope to show that the condition cannot be satisfied. It matters little in one respect: the investigation, if it is well conducted, starts in both cases in the same way, examining the hypothetical situation in which the condition is fulfilled, and shows only in its later course which hope is justified.

Compare FIGURES, 2. Compare also PAPPUS; an analysis which ends in disproving the proposed theorem, or in showing that the proposed "problem to find" has no solution, is actually a "reductio ad absurdum."

3. *Indirect proof.* The prime numbers, or primes, are the numbers 2, 3, 5, 7, 11, 13, 17, 19, 23, 29, 31, 37, . . . which cannot be resolved into smaller factors, although they are greater than 1. (The last clause excludes the number 1 which, obviously, cannot be resolved into smaller factors, but has a different nature and should not be counted as a prime.) The primes are the "ultimate elements" into which all integers (greater than 1) can be decomposed. For instance,

$$630 = 2 \cdot 3 \cdot 3 \cdot 5 \cdot 7$$

is decomposed into a product of five primes.

Is the series of primes infinite or does it end somewhere? It is natural to suspect that the series of primes never ends. If it ended somewhere, all integers could be decomposed into a finite number of ultimate elements and the world would appear "too poor" in a manner of speaking. Thus arises the problem of proving the existence of an infinity of prime numbers.

This problem is very different from elementary mathematical problems of the usual kind and appears at first inaccessible. Yet, as we said, it is extremely unlikely that there should be a last prime, say P. Why is it so unlikely?

Let us face squarely the unlikely situation in which, hypothetically, supposedly, allegedly, there is a last prime P. Then we could write down the complete series of primes 2, 3, 5, 7, 11, . . . P. Why is this so unlikely? What is wrong with it? Can we point out anything that is definitely wrong? Indeed, we can. We can construct the number

$$Q = (2 \cdot 3 \cdot 5 \cdot 7 \cdot 11 \ldots P) + 1.$$

This number Q is greater than P and therefore, allegedly, Q cannot be a prime. Consequently, Q must be divisible by a prime. Now, all primes at our disposal are, supposedly, the numbers 2, 3, 5, . . . P but Q, divided by any of these numbers, leaves the rest 1; and so Q is not divisible by any of the primes mentioned which are, hypothetically, all the primes. Now, there is something that is definitely wrong; Q must be either a prime or it must be divisible by some prime. Starting from the supposition that there is a last prime P we have been led to a manifest absurdity. How can we explain this? Our original supposition must be wrong; there cannot be a last prime P. And so we have succeeded in proving that the series of prime numbers never ends.

Our proof is a typical indirect proof. (It is a famous proof too, due to Euclid; see Proposition 20 of Book IX of the Elements.)

We have established our theorem (that the series of primes never ends) by disproving its contradictory opposite (that the series of primes ends somewhere) which we have disproved by deducing from it a manifest absurdity. Thus we have combined indirect proof with "reductio ad absurdum"; this combination is also very typical.

4. *Objections.* The procedures which we are studying encountered considerable opposition. Many objections have been raised which are, possibly, only various forms of the same fundamental objection. We discuss here a "practical" form of the objection, which is on our level.

To find a not obvious proof is a considerable intellectual achievement but to learn such a proof, or even to understand it thoroughly costs also a certain amount of mental effort. Naturally enough, we wish to retain some benefit from our effort, and, of course, what we retain in our memory should be true and correct and not false or absurd.

But it seems difficult to retain something true from a "reductio ad absurdum." The procedure starts from a false assumption and derives from it consequences which are equally, but perhaps more visibly, false till it reaches a last consequence which is manifestly false. If we do not wish to store falsehoods in our memory we should forget everything as quickly as possible which is, however, not feasible because all points must be remembered sharply and correctly during our study of the proof.

The objection to indirect proofs can be now stated very briefly. Listening to such a proof, we are obliged to focus our attention all the time upon a false assumption which we should forget and not upon the true theorem which we should retain.

If we wish to judge correctly of the merits of these objections, we should distinguish between two uses of the "reductio ad absurdum," as a tool of research and as a means of exposition, and make the same distinction concerning the indirect proof.

It must be confessed that "reductio ad absurdum" as a means of exposition is not an unmixed blessing. Such a "reductio," especially if it is long, may become very painful indeed for the reader or listener. All the derivations which we examine in succession are correct but all the situations which we have to face are impossible. Even the verbal expression may become tedious if it insists, as it should, on emphasizing that everything is based on an initial assumption; the words "hypothetically," "supposedly," "allegedly" must recur incessantly, or some other device must be applied continually. We wish to reject and forget the situation as impossible but we have to retain and examine it as the basis for the next step, and this inner discord may become unbearable in the long run.

Yet it would be foolish to repudiate "reductio ad absurdum" as a tool of discovery. It may present itself naturally and bring a decision when all other means seem to be exhausted as the foregoing examples may show.

We need some experience to perceive that there is no essential opposition between our two contentions. Experience shows that usually there is little difficulty in converting an indirect proof into a direct proof, or in rearranging a proof found by a long "reductio ad absurdum" into a more pleasant form from which the "reductio ad absurdum" may even completely disappear (or, after due preparation, it may be compressed into a few striking sentences).

In short, if we wish to make full use of our capacities, we should be familiar both with "reductio ad absurdum" and with indirect proof. When, however, we have succeeded in deriving a result by either of these methods we should not fail to look back at the solution and ask: *Can you derive the result differently?*

Let us illustrate by examples what we have said.

5. *Rearranging a reductio ad absurdum.* We look back at the reasoning presented under 1. The reductio ad absurdum started from a situation which, eventually, turned out to be impossible. Let us however carve out a part of the argument which is independent of the initial false assumption and contains positive information. Reconsidering what we have done, we may perceive that this much is doubtless true: If a set of numbers with one

or two digits is written so that each of the ten figures occurs just once, then the sum of the set is of the form

$$10t + (45 - t) = 9(t + 5).$$

Thus, this sum is divisible by 9. The proposed puzzle demands however that this sum should be 100. Is this possible? No, it is not, since 100 is not divisible by 9.

The "reductio ad absurdum" which led to the discovery of the argument vanished from our new presentation.

By the way, a reader acquainted with the procedure of "casting out nines" can see now the whole argument at a glance.

6. *Converting an indirect proof.* We look back at the reasoning presented under 3. Reconsidering carefully what we have done, we may find elements of the argument which are independent of any false assumption, yet the best clue comes from a reconsideration of the meaning of the original problem itself.

What do we mean by saying that the series of primes never ends? Evidently, just this: when we have ascertained any finite set of primes as 2, 3, 5, 7, 11, . . . P, where P is the last prime hitherto found, there is always one more prime. Thus, what must we do to prove the existence of an infinity of primes? We have to point out a way of finding a prime different from all primes hitherto found. Thus, our "problem to prove" is in fact reduced to a "problem to find": *Being given the primes 2, 3, 5, . . . P, find a new prime N different from all the given primes.*

Having restated our original problem in this new form, we have taken the main step. It is relatively easy now to see how to use the essential parts of our former argument for the new purpose. In fact, the number

$$Q = (2 \cdot 3 \cdot 5 \cdot 7 \cdot 11 \ldots P) + 1$$

is certainly divisible by a prime. Let us take—this is the

idea—any prime divisor of Q (for instance, the smallest one) for N. (Of course, if Q happens to be a prime, then $N = Q$.) Obviously, Q divided by any of the primes 2, 3, 5, . . . P leaves the remainder 1 and, therefore, none of these numbers can be N which is a divisor of Q. But that is all we need: N is a prime, and different from all hitherto found primes 2, 3, 5, 7, 11, . . . P.

This proof gives a definite procedure of prolonging again and again the series of primes, without limit. Nothing is indirect in it, no impossible situation needs to be considered. Yet, fundamentally, it is the same as our former indirect proof which we have succeeded in converting.

Redundant. See CONDITION.

Routine problem may be called the problem to solve the equation $x^2 - 3x + 2 = 0$ if the solution of the general quadratic equation was explained and illustrated before so that the student has nothing to do but to substitute the numbers -3 and 2 for certain letters which appear in the general solution. Even if the quadratic equation was not solved generally in "letters" but half a dozen similar quadratic equations with numerical coefficients were solved just before, the problem should be called a "routine problem." In general, a problem is a "routine problem" if it can be solved either by substituting special data into a formerly solved general problem, or by following step by step, without any trace of originality, some well-worn conspicuous example. Setting a routine problem, the teacher thrusts under the nose of the student an immediate and decisive answer to the question: *Do you know a related problem?* Thus, the student needs nothing but a little care and patience in following a cut-and-dried precept, and he has no opportunity to use his judgment or his inventive faculties.

Routine problems, even many routine problems, may be necessary in teaching mathematics but to make the students do no other kind is inexcusable. Teaching the mechanical performance of routine mathematical operations and nothing else is well under the level of the cookbook because kitchen recipes do leave something to the imagination and judgment of the cook but mathematical recipes do not.

Rules of discovery. The first rule of discovery is to have brains and good luck. The second rule of discovery is to sit tight and wait till you get a bright idea.

It may be good to be reminded somewhat rudely that certain aspirations are hopeless. Infallible rules of discovery leading to the solution of all possible mathematical problems would be more desirable than the philosophers' stone, vainly sought by the alchemists. Such rules would work magic; but there is no such thing as magic. To find unfailing rules applicable to all sorts of problems is an old philosophical dream; but this dream will never be more than a dream.

A reasonable sort of heuristic cannot aim at unfailing rules; but it may endeavor to study procedures (mental operations, moves, steps) which are typically useful in solving problems. Such procedures are practiced by every sane person sufficiently interested in his problem. They are hinted by certain stereotyped questions and suggestions which intelligent people put to themselves and intelligent teachers to their students. A collection of such questions and suggestions, stated with sufficient generality and neatly ordered, may be less desirable than the philosophers' stone but can be provided. The list we study provides such a collection.

Rules of style. The first rule of style is to have something to say. The second rule of style is to control yourself when, by chance, you have two things to say; say first one, then the other, not both at the same time.

Rules of teaching. The first rule of teaching is to know what you are supposed to teach. The second rule of teaching is to know a little more than what you are supposed to teach.

First things come first. The author of this book does not think that all rules of conduct for teachers are completely useless; otherwise, he would not have dared to write a whole book about the conduct of teachers and students. Yet it should not be forgotten that a teacher of mathematics should know some mathematics, and that a teacher wishing to impart the right attitude of mind toward problems to his students should have acquired that attitude himself.

Separate the various parts of the condition. Our first duty is to understand the problem. Having understood the problem as a whole, we go into detail. We consider its principal parts, the unknown, the data, the condition, each by itself. When we have these parts well in mind but no particularly helpful idea has yet occurred to us, we go into further detail. We consider the various data, each datum by itself. Having understood the condition as a whole, we *separate its various parts,* and we consider each part by itself.

We see now the role of the suggestion that we have to discuss here. It tends to provoke a step that we have to take when we are trying to see the problem distinctly and have to go into finer and finer detail. It is a step in DECOMPOSING AND RECOMBINING.

Separate the various parts of the condition. Can you write them down? We often have opportunity to ask this question when we are SETTING UP EQUATIONS.

Setting up equations is like translation from one language into another (NOTATION, 1). This comparison, used by Newton in his *Arithmetica Universalis,* may help to clarify the nature of certain difficulties often felt both by students and by teachers.

1. To set up equations means to express in mathematical symbols a condition that is stated in words; it is translation from ordinary language into the language of mathematical formulas. The difficulties which we may have in setting up equations are difficulties of translation.

In order to translate a sentence from English into French two things are necessary. First, we must understand thoroughly the English sentence. Second, we must be familiar with the forms of expression peculiar to the French language. The situation is very similar when we attempt to express in mathematical symbols a condition proposed in words. First, we must understand thoroughly the condition. Second, we must be familiar with the forms of mathematical expression.

An English sentence is relatively easy to translate into French if it can be translated word for word. But there are English idioms which cannot be translated into French word for word. If our sentence contains such idioms, the translation becomes difficult; we have to pay less attention to the separate words, and more attention to the whole meaning; before translating the sentence, we may have to rearrange it.

It is very much the same in setting up equations. In easy cases, the verbal statement splits almost automatically into successive parts, each of which can be immediately written down in mathematical symbols. In more difficult cases, the condition has parts which cannot be immediately translated into mathematical symbols. If this is so, we must pay less attention to the verbal statement, and concentrate more upon the meaning. Before we start writing formulas, we may have to rearrange the condition, and we should keep an eye on the resources of mathematical notation while doing so.

In all cases, easy or difficult, we have to understand the condition, to *separate the various parts of the condition,* and to ask: *Can you write them down?* In easy cases, we succeed without hesitation in dividing the condition into parts that can be written down in mathematical symbols; in difficult cases, the appropriate division of the condition is less obvious.

The foregoing explanation should be read again after the study of the following examples.

2. *Find two quantities whose sum is* 78 *and whose product is* 1296.

We divide the page by a vertical line. On one side, we write the verbal statement split into appropriate parts. On the other side, we write algebraic signs, opposite to the corresponding part of the verbal statement. The original is on the left, the translation into symbols on the right.

Stating the problem

in English	in algebraic language
Find two quantities	$x, \quad y$
whose sum is 78 and	$x + y = 78$
whose product is 1296	$xy = 1296.$

In this case, the verbal statement splits almost automatically into successive parts, each of which can be immediately written down in mathematical symbols.

3. *Find the breadth and the height of a right prism with square base, being given the volume,* 63 *cu. in., and the area of the surface,* 102 *sq. in.*

What are the unknowns? The side of the base, say x, and the altitude of the prism, say y.

What are the data? The volume, 63, and the area, 102.

What is the condition? The prism whose base is a square with side x and whose altitude is y must have the volume 63 and the area 102.

Separate the various parts of the condition. There are two parts, one concerned with the volume, the other with the area.

We can scarcely hesitate in dividing the whole condition just in these two parts; but we cannot write down these parts "immediately." We must know how to calculate the volume and the various parts of the area. Yet, if we know that much geometry, we can easily restate both parts of the condition so that the translation into equations is feasible. We write on the left hand side of the page an essentially rearranged and expanded statement of the problem, ready for translation into algebraic language.

Of a right prism with square base	
find the side of the base	x
and the altitude.	y
First. The volume is given.	63
The area of the base which is a square with side x	x^2
and the altitude	y
determine the volume which is their product.	$x^2y = 63$
Second. The area of the surface is given.	102
The surface consists of two squares with side x	$2x^2$
and of four rectangles, each with base x and altitude y,	$4xy$
whose sum is the area.	$2x^2 + 4xy = 102.$

4. *Being given the equation of a straight line and the*

coordinates of a point, find the point which is symmetrical to the given point with respect to the given straight line.

This is a problem of plane analytic geometry.

What is the unknown? A point, with coordinates, say, p, q.

What is given? The equation of a straight line, say $y = mx + n$, and a point with coordinates, say, a, b.

What is the condition? The points (a, b) and (p, q) are symmetrical to each other with respect to the line $y = mx + n$.

We now reach the essential difficulty which is to divide the condition into parts each of which can be expressed in the language of analytic geometry. The nature of this difficulty must be well understood. A decomposition of the condition into parts may be logically unobjectionable and nevertheless useless. What we need here is a decomposition into parts which are fit for analytic expression. In order to find such a decomposition we must *go back to the definition* of symmetry, but keep an eye on the resources of analytic geometry. What is meant by symmetry with respect to a straight line? What geometric relations can we express simply in analytic geometry? We concentrate upon the first question, but we should not forget the second. Thus, eventually, we may find the decomposition which we are going to state.

<table>
<tr><td>The given point
and the point required</td><td>(a, b)
(p, q)</td></tr>
<tr><td>are so related that
first, the line joining them
is perpendicular to the
given line, and</td><td>$$\frac{q - b}{p - a} = -\frac{1}{m}$$</td></tr>
<tr><td>second, the midpoint of
the line joining them lies
on the given line.</td><td>$$\frac{b + q}{2} = m\frac{a + p}{2} + n.$$</td></tr>
</table>

Signs of progress. As Columbus and his companions sailed westward across an unknown ocean they were cheered whenever they saw birds. They regarded a bird as a favorable sign, indicating the nearness of land. But in this they were repeatedly disappointed. They watched for other signs too. They thought that floating seaweed or low banks of cloud might indicate land, but they were again disappointed. One day, however, the signs multiplied. On Thursday, the 11th of October, 1492, "they saw sandpipers, and a green reed near the ship. Those of the caravel *Pinta* saw a cane and a pole, and they took up another small pole which appeared to have been worked by iron; also another bit of cane, a land-plant, and a small board. The crew of the caravel *Niña* also saw signs of land, and a small branch covered with berries. Everyone breathed afresh and rejoiced at these signs." And in fact the next day they sighted land, the first island of a New World.

Our undertaking may be important or unimportant, our problem of any kind—when we are working intensely, we watch eagerly for signs of progress as Columbus and his companions watched for signs of approaching land. We shall discuss a few examples in order to understand what can be reasonably regarded as a sign of approaching the solution.

1. *Examples.* I have a chess problem. I have to mate the black king in, say, two moves. On the chessboard there is a white knight, quite a distance from the black king, that is apparently superfluous. What is it good for? I am obliged to leave this question unanswered at first. Yet after various trials, I hit upon a new move and observe that it would bring that apparently superfluous white knight into play. This observation gives me a new hope. I regard it as a favorable sign: that new move has some chance to be the right one. Why?

In a well-constructed chess problem there is no superfluous piece. Therefore, we have to take into account all chessmen on the board; we have to *use all the data.* The correct solution does certainly use all the pieces, even that apparently superfluous white knight. In this last respect, the new move that I contemplate agrees with the correct move that I am supposed to find. The new move looks like the correct move; it might be the correct move.

It is interesting to consider a similar situation in a mathematical problem. My task is to express the area of a triangle in terms of its three sides, a, b, and c. I have already made some sort of plan. I know, more or less clearly, which geometrical connections I have to take into account and what sort of calculations I have to perform. Yet I am not quite sure whether my plan will work. If now, proceeding along the line prescribed by my plan, I observe that the quantity

$$\sqrt{b + c - a}$$

enters into the expression of the area I am about to construct, I have good reason to be cheered. Why?

In fact, it must be taken into account that the sum of any two sides of a triangle is greater than the third side. This involves a certain restriction. The given lengths, a, b, and c cannot be quite arbitrary; for instance, $b + c$ must be greater than a. This is an essential part of the condition, and we should *use the whole condition.* If $b + c$ is not greater than a the formula I seek is bound to become illusory. Now, the square root displayed above becomes imaginary if $b + c - a$ is negative—that is, if $b + c$ is less than a—and so the square root becomes unfit to represent a real quantity under just those circumstances under which the desired expression is bound to become illusory. Thus my formula, into which that square root enters, has an important property in common with the true formula for the area. My formula looks like the true formula; it might be the true formula.

Here is one more example. Some time ago, I wished to prove a theorem in solid geometry. Without much trouble I found a first remark that appeared to be pertinent; but then I got stuck. Something was missing to finish the proof. When I gave up that day I had a much clearer notion than at the outset how the proof should look, how the gap should be filled; but I was not able to fill it. The next day, after a good night's rest, I looked again into the question and soon hit upon an analogous theorem in plane geometry. In a flash I was convinced that now I had got hold of the solution and I had, I think, good reason too to be convinced. Why?

In fact, *analogy* is a great guide. The solution of a problem in solid geometry often depends on an analogous problem in plane geometry (see ANALOGY, 3-7). Thus, in my case, there was a chance from the outset that the desired proof would use as a lemma some theorem of plane geometry of the kind which actually came to my mind. "This theorem looks like the lemma I need; it might be the lemma I need"—such was my reasoning.

If Columbus and his men had taken the trouble to reason explicitly, they would have reasoned in some similar way. They knew how the sea looks near the shore. They knew that, more often than on the open sea, there are birds in the air, coming from the land, and objects floating in the water, detached from the seashore. Many of the men must have observed such things when from former voyages they had returned to their home port. The day before that memorable date on which they sighted the island of San Salvador, as the floating objects in the water became so frequent, they thought: "It looks as if we were approaching some land; we may be approaching some land" and "everyone breathed afresh and rejoiced at these signs."

2. *Heuristic character of signs of progress.* Let us insist upon a point which is perhaps already clear to everyone; but it is very important and, therefore, it should be completely clear.

The type of reasoning illustrated by the foregoing examples deserves to be noticed and taken into account seriously, although it yields only a plausible indication and not an unfailing certainty. Let us restate pedantically, at full length, in rather unnatural detail, one of these reasonings:

> If we are approaching land, we often see birds.
> Now we see birds.
> Therefore, probably, we are approaching land.

Without the word "probably" the conclusion would be an outright fallacy. In fact, Columbus and his companions saw birds many times but were disappointed later. Just once came the day on which they saw sandpipers followed by the day of discovery.

With the word "probably" the conclusion is reasonable and natural but by no means a proof, a demonstrative conclusion; it is only an indication, a heuristic suggestion. It would be a great mistake to forget that such a conclusion is only probable, and to regard it as certain. But to disregard such conclusions entirely would be a still greater mistake. If you take a heuristic conclusion as certain, you may be fooled and disappointed; but if you neglect heuristic conclusions altogether you will make no progress at all. The most important signs of progress are heuristic. Should we trust them? Should we follow them? Follow, but keep your eyes open. Trust but look. And never renounce your judgment.

3. *Clearly expressible signs.* We can look at the foregoing examples from another point of view.

In one of these examples, we regarded as a favorable sign that we succeeded in bringing into play a datum not used before (the white knight). We were quite right to so regard it. In fact, to solve a problem is, essentially, to *find the connection between the data and the unknown.* Moreover we should, at least in well-stated problems, *use all the data,* connect each of them with the unknown. Thus, bringing one more datum into play is quite properly felt as progress, as a step forward.

In another example, we regarded as a favorable sign that an essential clause of the condition was appropriately taken into account by our formula. We were quite right to so regard it. In fact, we should *use the whole condition.* Thus, taking into account one more clause of the condition is justly felt as progress, as a move in the right direction.

In still another example, we regarded as a favorable sign the emergence of a simpler analogous problem. This also is justified. Indeed, analogy is one of the main sources of invention. If other means fail, we should try to *imagine an analogous problem.* Therefore, if such a problem emerges spontaneously, by its own accord, we naturally feel elated; we feel that we are approaching the solution.

After these examples, we can now easily grasp the general idea. There are certain mental operations typically useful in solving problems. (The most usual operations of this kind are listed in this book.) If such a typical operation succeeds (if one more datum is connected with the unknown—one more clause of the condition is taken into account—a simpler analogous problem is introduced) its success is felt as a sign of progress. Having understood this essential point, we can express with some clearness the nature of still other signs of progress. All we have to do is to read down our list and look at the various questions and suggestions from our newly acquired point of view.

Thus, understanding clearly the nature of the unknown means progress. Clearly disposing the various data so that we can easily recall any one also means progress. Visualizing vividly the condition as a whole may mean an essential advance; and separating the condition into appropriate parts may be an important step forward. When we have found a figure that we can easily imagine, or a notation that we can easily retain, we can reasonably believe that we have made some progress. Recalling a *problem related to ours and solved before* may be a decisive move in the right direction.

And so on, and so forth. To each mental operation clearly conceived corresponds a certain sign clearly expressible. Our list, appropriately read, lists also signs of progress.

Now, the questions and suggestions of our list are simple, obvious, just plain common sense. This has been said repeatedly and the same can be said of the connected signs of progress we discuss here. To read such signs no occult science is needed, only a little common sense and, of course, a little experience.

4. *Less clearly expressible signs.* When we work intently, we feel keenly the pace of our progress: when it is rapid we are elated; when it is slow we are depressed. We feel such differences quite clearly without being able to point out any distinct sign. Moods, feelings, general aspects of the situation serve to indicate our progress. They are not easy to express. "It looks good to me," or "It is not so good," say the unsophisticated. More sophisticated people express themselves with some nuance: "This is a well-balanced plan," or "No, something is still

lacking and that spoils the harmony." Yet behind primitive or vague expressions there is an unmistakable feeling which we follow with confidence and which leads us frequently in the right direction. If such feeling is very strong and emerges suddenly, we speak of inspiration. People usually cannot doubt their inspirations and are sometimes fooled by them. In fact, we should treat guiding feelings and inspirations just as we treat the more clearly expressible signs of progress which we have considered before. Trust, but keep your eyes open.

Always follow your inspiration—with a grain of doubt.

[What is the nature of those guiding feelings? Is there some less vague meaning behind words of such aesthetic nuances as "well-balanced," or "harmonious"? These questions may be more speculative than practical, but the present context indicates answers which perhaps deserve to be stated: Since the more clearly expressible signs of progress are connected with the success or failure of certain rather definite mental operations, we may suspect that our less clearly expressible guiding feelings may be similarly connected with other, more obscure, mental activities—perhaps with activities whose nature is more "psychological" and less "logical."]

5. *How signs help.* I have a plan. I see pretty clearly where I should begin and which steps I should take first. Yet I do not quite see the lay-out of the road farther on; I am not quite certain that my plan will work; and, in any case, I have still a long way to go. Therefore, I start out cautiously in the direction indicated by my plan and keep a lookout for signs of progress. If the signs are rare or indistinct, I become more hesitant. And if for a long time they fail to appear altogether, I may lose courage, turn back, and try another road. On the other hand, if the signs become more frequent as I proceed, if they multiply, my hesitation fades, my spirits rise, and I move

with increasing confidence, just as Columbus and his companions did before sighting the island of San Salvador.

Signs may guide our acts. Their absence may warn us of a blind alley and save us time and useless exertion; their presence may cause us to concentrate our effort upon the right spot.

Yet signs may also be deceptive. I once abandoned a certain path for lack of signs, but a man who came after me and followed that path a little farther made an important discovery—to my great annoyance and long-lasting regret. He not only had more perseverance than I did but he also read correctly a certain sign which I had failed to notice. Again, I may follow a road cheerfully, encouraged by favorable signs, and run against an unsuspected and insurmountable obstacle.

Yes, signs may misguide us in any single case, but they guide us right in the majority of them. A hunter may misinterpret now and then the traces of his game but he must be right on the average, otherwise he could not make a living by hunting.

It takes experience to interpret the signs correctly. Some of Columbus's companions certainly knew by experience how the sea looks near the shore and so they were able to read the signs which suggested that they were approaching land. The expert knows by experience how the situation looks and feels when the solution is near and so he is able to read the signs which indicate that he is approaching it. The expert knows more signs than the inexperienced, and he knows them better; his main advantage may consist in such knowledge. An expert hunter notices traces of game and appraises even their freshness or staleness where the inexperienced one is unable to see anything.

The main advantage of the exceptionally talented may

consist in a sort of extraordinary mental sensibility. With exquisite sensibility, he feels subtle signs of progress or notices their absence where the less talented are unable to perceive a difference.

[6. *Heuristic syllogism.* In section 2 we came across a mode of heuristic reasoning that deserves further consideration and a technical term. We begin by restating that reasoning in the following form:

> If we are approaching land, we often see birds.
> Now we see birds.
> ___
> Therefore, it becomes more credible that we are approaching land.

The two statements above the horizontal line may be called the *premises,* the statement under the line, the *conclusion.* And the whole pattern of reasoning may be termed a *heuristic syllogism.*

The premises are stated here in the same form as in section 2, but the conclusion is more carefully worded. An essential circumstance is better emphasized. Columbus and his men conjectured from the beginning that they would eventually find land sailing westward; and they must have given some credence to this conjecture, otherwise they would not have started out at all. As they proceeded, they related every incident, major or minor, to their dominating question: "Are we approaching land?" Their confidence rose and fell as events occurred or failed to occur, and each man's beliefs fluctuated more or less differently according to his background and character. The whole dramatic tension of the voyage is due to such fluctuations of confidence.

The heuristic syllogism quoted exhibits a reasonable ground for a change in the level of confidence. To occasion such changes is the essential role of this kind of reasoning and this point is better expressed by the wording given here than by the one in section 2.

The general pattern suggested by our example can be exhibited thus:

If A is true, then B is also true, as we know.
Now, it turns out that B is true.

Therefore, A becomes more credible.

Still shorter:

$$\frac{\text{If } A \text{ then } B \qquad B \text{ true}}{A \text{ more credible}}$$

In this schematic statement the horizontal line stands for the word "therefore" and expresses the implication, the essential link between the premises and the conclusion.]

[7. *Nature of plausible reasoning.* In this little book we are discussing a philosophical question. We discuss it as practically and informally and as far from high-brow modes of expression as we can, but nevertheless our subject is philosophical. It is concerned with the nature of heuristic reasoning and, by extension, with a kind of reasoning which is nondemonstrative although important and which we shall call, for lack of a better term, *plausible* reasoning.

The signs that convince the inventor that his idea is good, the indications that guide us in our everyday affairs, the circumstantial evidence of the lawyer, the inductive evidence of the scientist, statistical evidence invoked in many and diverse subjects—all these kinds of evidence agree in two essential points. First, they do not have the certainty of a strict demonstration. Second, they are useful in acquiring essentially new knowledge, and even indispensable to any not purely mathematical or logical knowledge, to any knowledge concerned with the physical world. We could call the reasoning that underlies this kind of evidence "heuristic reasoning" or "inductive reasoning" or (if we wish to avoid stretching the meaning of existing terms) "plausible reasoning." We accept here the last term.

The heuristic syllogism introduced in the foregoing may be regarded as the simplest and most widespread pattern of plausible reasoning. It reminds us of a classical pattern of demonstrative reasoning, of the so-called "modus tollens of hypothetical syllogism." We exhibit here both patterns side by side:

Demonstrative	*Heuristic*
If A then B	If A then B
B false	B true
A false	A more credible

The comparison of these patterns may be instructive. It may grant us an insight, not easily obtainable elsewhere, into the nature of plausible (heuristic, inductive) reasoning.

Both patterns have the same first premise:

If A then B.

They differ in the second premise. The statements:

B false $\qquad\qquad$ B true

are exactly opposite to each other but they are of "similar logical nature," they are on the same "logical level." The great difference arises after the premises. The conclusions

A false $\qquad\qquad$ A more credible

are on different logical levels and their relations to their respective premises are of a different logical nature.

The conclusion of the demonstrative syllogism is of the

same logical nature as the premises. Moreover, this conclusion is fully expressed and is fully supported by the premises. If my neighbor and I agree to accept the premises, we cannot reasonably disagree about accepting also the conclusion, however different our tastes or other convictions may be.

The conclusion of the heuristic syllogism differs from the premises in its logical nature; it is more vague, not so sharp, less fully expressed. This conclusion is comparable to a force, has direction and magnitude. It pushes us in a certain direction: *A* becomes *more* credible. The conclusion also has a certain strength: *A* may become *much more* credible, or *just a little more* credible. The conclusion is not fully expressed and is not fully supported by the premises. *The direction is expressed and is implied by the premises, the magnitude is not.* For any reasonable person, the premises involve that *A* becomes more credible (certainly not less credible). Yet my neighbor and I can honestly disagree *how much* more credible *A* becomes, since our temperaments, our backgrounds, and our unstated reasons may be different.

In the demonstrative syllogism the premises constitute a *full basis* on which the conclusion rests. If both premises stand, the conclusion stands too. If we receive some new information that does not change our belief in the premises, it cannot change our belief in the conclusion.

In the heuristic syllogism the premises constitute only one *part of the basis* on which the conclusion rests, the fully expressed, the "visible" part of the basis; there is an unexpressed, invisible part, formed by something else, by inarticulate feelings perhaps, or by unstated reasons. In fact, it can happen that we receive some new information that leaves our belief in both premises completely intact, but influences the trust we put in *A* in a way just opposite to that expressed in the conclusion. To find *A* more plausible on the ground of the premises of our heuristic syllogism is only reasonable. Yet tomorrow I may find grounds, not interfering at all with these premises, that make A appear less plausible, or even definitively refute it. The conclusion may be shaken and even overturned completely by commotions in the invisible parts of its foundation, although the premises, the visible part, stand quite firm.

These remarks seem to make somewhat more understandable the nature of heuristic, inductive, and other sorts of not demonstrative plausible reasoning, which appear so baffling and elusive when approached from the point of view of purely demonstrative logic. Many more concrete examples, the consideration of other kinds of heuristic syllogism, and an investigation of the concept of probability and other allied concepts seem to be necessary to complete the approach here chosen; *cf.* the author's *Mathematics and Plausible Reasoning.*]

Heuristic reasons are important although they prove nothing. To clarify our heuristic reasons is also important although behind any reason clarified there are many others that remain obscure and are perhaps still more important.

Specialization is passing from the consideration of a given set of objects to that of a smaller set, or of just one object, contained in the given set. Specialization is often useful in the solution of problems.

1. *Example.* In a triangle, let r be the radius of the inscribed circle, R the radius of the circumscribed circle, and H the longest altitude. Then

$$r + R \leqq H.$$

We have to prove (or disprove) this theorem[9]; we have a "problem to prove."

[9] The *American Mathematical Monthly,* vol. 50 (1943), p. 124 and vol. 51 (1944), pp. 234-236.

The proposed theorem is of an unusual sort. We can scarcely remember any theorem about triangles with a similar conclusion. If nothing else occurs to us, we may test some *special case* of this unfamiliar assertion. Now, the best known special triangle is the equilateral triangle for which

$$r = \frac{H}{3} \qquad R = \frac{2H}{3}$$

so that, in this case, the assertion is correct.

If no other idea presents itself, we may test the *more extended special case* of isosceles triangles. The form of an isosceles triangle varies with the angle at the vertex and there are two extreme (or limiting) cases, the one in which the angle at the vertex becomes 0°, and the other in which it becomes 180°. In the first extreme case the base of the isosceles triangle vanishes and visibly

$$r = 0 \qquad R = \frac{1}{2}H$$

thus, the assertion is verified. In the second limiting case, however, all three heights vanish and

$$r = 0 \qquad R = \infty \qquad H = 0.$$

The assertion is not verified. We have proved that the proposed theorem is false, and so we have solved our problem.

By the way, it is clear that the assertion is also false for very flat isosceles triangles whose angle at the vertex is nearly 180° so that we may "officially" disregard the extreme cases whose consideration may appear as not quite "orthodox."

2. "L'exception confirme la règle." "The exception proves the rule." We must take this widely known saying as a joke, laughing at the laxity of a certain sort of logic. If we take matters seriously, one exception is enough, of course, to refute irrefragably any would-be rule or gen-

eral statement. The most usual and, in some respects, the best method to refute such a statement consists precisely in exhibiting an object that does not comply with it; such an object is called a *counter-example* by certain writers.

The allegedly general statement is concerned with a certain set of objects; in order to refute the statement we *specialize,* we pick out from the set an object that does not comply with it. The foregoing example (under 1) shows how it is done. We may examine at first any simple special case, that is, any object chosen more or less at random which we can easily test. If the test shows that the case is not in accordance with the general statement, the statement is refuted and our task finished. If, however, the object examined complies with the statement we may possibly derive some hint from its examination. We may receive the impression that the statement could be true, after all, and some suggestion in which direction we should seek the proof. Or, we may receive, as in our example under 1, some suggestion in which direction we should seek the counter-example, that is, which other special cases should we test. We may modify the case we have just examined, vary it, investigate some more extended special case, look around for extreme cases, as exemplified under 1.

Extreme cases are particularly instructive. If a general statement is supposed to apply to all mammals it must apply even to such an unusual mammal as the whale. Let us not forget this extreme case of the whale. Examining it, we may refute the general statement; there is a good chance for that, since such extreme cases are apt to be overlooked by the inventors of generalizations. If, however, we find that the general statement is verified even in the extreme case, the inductive evidence derived from this verification will be strong, just because the prospect of refutation was strong. Thus, we are tempted to reshape the saying from which we started: "Prospective exceptions test the rule."

3. *Example.* Given the speeds of two ships and their positions at a certain moment; each ship steers a rectilinear course with constant speed. Find the distance of the two ships when they are nearest to each other.

What is the unknown? The shortest distance between two moving bodies. The bodies have to be considered as material points.

What are the data? The initial positions of the moving material points, and the speed of each. These speeds are constant in amount and direction.

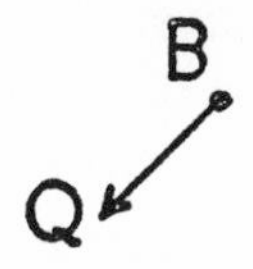

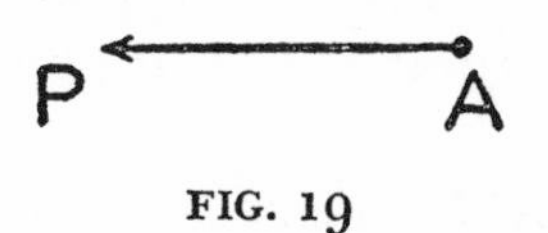

FIG. 19

What is the condition? The distance has to be ascertained when it is the shortest, that is, at the moment when the two moving points (ships) are nearest to each other.

Draw a figure. Introduce suitable notation. In Fig. 19, the points A and B mark the given initial positions of the two ships. The directed line-segments (vectors) AP and BQ represent the given speeds so that the first ship proceeds along the straight line through the points A and P, and covers the distance AP in unit time. The second ship travels similarly along the straight line BQ.

What is the unknown? The shortest distance of the two ships, the one traveling along AP and the other along BQ.

It is clear by now what we should find; yet, if we wish to use only elementary means, we may be still in the dark how we should find it. The problem is not too easy and its difficulty has some peculiar nuance which we may try to express by saying that "there is too much variety." The initial positions, A and B, and the speeds, AP and BQ, can be given in various ways; in fact, the four points A, B, P, Q may be chosen arbitrarily. Now, whatever the data may be, the required solution must apply and we do not see yet how to fit the same solution to all these possibilities. Out of such feeling of "too much variety" this question and answer may eventually emerge:

Could you imagine a more accessible related problem? A more special problem? Of course, there is the extreme special case in which one of the speeds vanishes. Yes, the ship in B may lay at anchor, Q may coincide with B. The shortest distance from the ship at rest to the moving ship is the perpendicular to the straight line along which the latter moves.

4. If the foregoing idea emerges with the premonition that there is more ahead and with the feeling that that extreme special case (which could appear as too simple to be relevant) has some role to play—then it is a bright idea indeed.

Here is a problem related to yours, that specialized problem you just solved. *Could you use it? Could you use its result? Should you introduce some auxiliary element in order to make its use possible?* It should be used, but how? How could the result of the case in which B is at rest be used in the case in which B is moving? Rest is a *special case* of motion. And motion is relative—and, therefore, whatever the given velocity of B may be I can consider B as being at rest! Here is the idea more clearly: If I impart to the whole system, consisting of both ships, the same uniform motion, the relative positions do not change, the relative distances remain the same, and so does especially the shortest relative distance of the two ships required by the problem. Now, I can impart a motion that reduces the speed of one of the ships to zero,

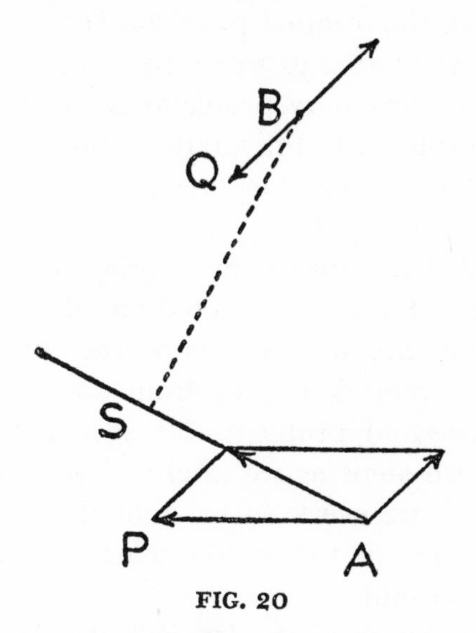

FIG. 20

and so reduces the general case of the problem to the special case just solved. Let me add a velocity, opposite to BQ but of the same amount, both to BQ and to AP. This is the auxiliary element that makes the use of the special result possible.

See Fig. 20 for the construction of the shortest distance, BS.

5. The foregoing solution (under 3, 4) has a logical pattern that deserves to be analyzed and remembered.

In order to solve our original problem (under 3, first lines) we have solved first another problem which we may call appropriately the auxiliary problem (under 3, last lines). This auxiliary problem is a special case of the original problem (the extreme special case in which one of the two ships is at rest). The original problem was proposed, the auxiliary problem invented in the course of the solution. The original problem looked hard, the solution of the auxiliary problem was immediate. The auxiliary problem was, as a special case, in fact much *less ambitious* than the original problem. How is it then possible that we were able to solve the original problem on the basis of the auxiliary problem? Because in reducing the original problem to the auxiliary problem, we added a substantial supplementary remark (on relativity of motion).

We succeeded in solving our original problem thanks to two remarks. First, we invented an advantageous auxiliary problem. Second, we discovered an appropriate supplementary remark to pass from the auxiliary problem to the original problem. We solved the proposed problem in two steps as we might cross a creek in two steps provided we were lucky enough to discover an appropriate stone in the middle which could serve as a momentary foothold.

To sum up, we used the less difficult, less ambitious, special, auxiliary problem as a *stepping stone* in solving the more difficult, more ambitious, general, original problem.

6. Specialization has many other uses which we cannot discuss here. It may be just mentioned that it can be useful in testing the solution (CAN YOU CHECK THE RESULT? 2).

A somewhat primitive kind of specialization is often useful to the teacher. It consists in giving some *concrete interpretation* to the abstract mathematical elements of the problem. For instance, if there is a rectangular parallelepiped in the problem, the teacher may take the classroom in which he talks as example (section **8**). In solid analytic geometry, a corner of the classroom may serve as the origin of coordinates, the floor and two walls as coordinate planes, two horizontal edges of the room and one vertical edge as coordinate axes. Explaining the notion of a surface of revolution, the teacher draws a curve with chalk on the door and opens it slowly. These are certainly simple tricks but nothing should be omitted that has some chance to bring home mathematics to the students: Mathematics being a very abstract science should be presented very concretely.

Subconscious work. One evening I wished to discuss with a friend a certain author but I could not remember the author's name. I was annoyed, because I remembered fairly well one of his stories. I remembered also some story about the author himself which I wanted to tell; I remembered, in fact, everything except the name. Repeatedly, I tried to recollect that name but all in vain. The next morning, as soon as I thought of the annoyance of the evening before, the name occurred to me without any effort.

The reader, very likely, remembers some similar experience of his own. And, if he is a passionate problem-solver, he has probably had some similar experience with problems. It often happens that you have no success at all with a problem; you work very hard yet without finding anything. But when you come back to the problem after a night's rest, or a few days' interruption, a bright idea appears and you solve the problem easily. The nature of the problem matters little; a forgotten word, a difficult word from a crossword-puzzle, the beginning of an annoying letter, or the solution of a mathematical problem may occur in this way.

Such happenings give the impression of *subconscious work*. The fact is that a problem, after prolonged absence, may return into consciousness essentially clarified, much nearer to its solution than it was when it dropped out of consciousness. Who clarified it, who brought it nearer to the solution? Obviously, oneself, working at it *subconsciously*. It is difficult to give any other answer; although psychologists have discovered the beginnings of another answer which may turn out some day to be more satisfactory.

Whatever may or may not be the merits of the theory of subconscious work, it is certain that there is a limit beyond which we should not force the conscious reflection. There are certain moments in which it is better to leave the problem alone for a while. "Take counsel of your pillow" is an old piece of advice. Allowing an interval of rest to the problem and to ourselves, we may obtain more tomorrow with less effort. "If today will not, tomorrow may" is another old saying. But it is desirable not to set aside a problem to which we wish to come back later without the impression of some achievement; at least some little point should be settled, some aspect of the question somewhat elucidated when we quit working.

Only such problems come back improved whose solution we passionately desire, or for which we have worked with great tension; conscious effort and tension seem to be necessary to set the subconscious work going. At any rate, it would be too easy if it were not so; we could solve difficult problems just by sleeping and waiting for a bright idea.

Past ages regarded a sudden good idea as an inspiration, a gift of the gods. You must deserve such a gift by work, or at least by a fervent wish.[10]

Symmetry has two meanings, a more usual, particular, geometric meaning, and a less usual, general, logical meaning.

Elementary solid geometry considers two kinds of symmetry, symmetry with respect to a plane (called plane of symmetry), and symmetry with respect to a point (called center of symmetry). The human body appears to be fairly symmetrical but in fact it is not; many interior organs are quite unsymmetrically disposed. A statue may be completely symmetrical with respect to a vertical plane so that its two halves appear completely "interchangeable."

In a more general acceptance of the word, a whole is termed symmetric if it has interchangeable parts. There are many kinds of symmetry; they differ in the number of interchangeable parts, and in the operations which exchange the parts. Thus, a cube has high symmetry; its 6 faces are interchangeable with each other, and so are its 8 vertices, and so are its 12 edges. The expression

$$yz + zx + xy$$

is symmetric; any two of the three letters x, y, z can be interchanged without changing the expression.

Symmetry, in a general sense, is important for our subject. If a problem is symmetric in some ways we may derive some profit from noticing its interchangeable parts and it often pays to treat those parts which play the same role in the same fashion (see AUXILIARY ELEMENTS, 3).

[10] For an all-round discussion of "unconscious thinking" see Jacques Hadamard, *The Psychology of Invention in the Mathematical Field.*

Try to treat symmetrically what is symmetrical, and do not destroy wantonly any natural symmetry. However, we are sometimes compelled to treat unsymmetrically what is naturally symmetrical. A pair of gloves is certainly symmetrical; nevertheless, nobody handles the pair quite symmetrically, nobody puts on both gloves at the same time, but one after the other.

Symmetry may also be useful in checking results; see section **14.**

Terms, old and new, describing the activity of solving problems are often ambiguous. The activity itself is familiar to everybody and it is often discussed but, as other mental activities, it is difficult to describe. In the absence of a systematic study there are no technical terms to describe it, and certain usual half-technical terms often add to the confusion because they are used in different meanings by different authors.

The following short list includes a few new terms used and a few old terms avoided in the present study, and also some old terms retained despite their ambiguity.

The reader may be confused by the following discussion of terminology unless his notions are well anchored in examples.

1. *Analysis* is neatly defined by PAPPUS, and it is a useful term, describing a typical way of devising a plan, starting from the unknown (or the conclusion) and working backwards, toward the data (or the hypothesis). Unfortunately, the word has acquired very different meanings (for instance, of mathematical, chemical, logical analysis) and therefore, it is regretfully avoided in the present study.

2. *Condition* links the unknown of a "problem to find" to the data (see PROBLEMS TO FIND, PROBLEMS TO PROVE, 3). In this meaning, it is a clear, useful and un-

avoidable term. It is often necessary to decompose the condition into several parts [into parts (I) and (II) in the examples DECOMPOSING AND RECOMBINING, 7, 8]. Now, each part of *the* condition is usually called *a* condition. This ambiguity which is sometimes embarrassing could be easily avoided by introducing some technical term to denote the parts of the whole condition; for instance, such a part could be called a "clause."

3. *Hypothesis* denotes an essential part of a mathematical theorem of the more usual kind (see PROBLEMS TO FIND, PROBLEMS TO PROVE, 4). The term, in this meaning, is perfectly clear and satisfactory. The difficulty is that each part of *the* hypothesis is also called *a* hypothesis so that the hypothes*is* may consist of several hypothes*es*. The remedy would be to call each part of the whole hypothesis a "clause," or something similar. (Compare the foregoing remark on "condition.")

4. *Principal parts* of a problem are defined in PROBLEMS TO FIND, PROBLEMS TO PROVE, 3, 4.

5. *Problem to find, problem to prove* are a pair of new terms, introduced regretfully to replace historical terms whose meaning, however, is confused beyond redemption by current usage. In Latin versions of Greek mathematical texts, the common name for both kinds of problems is "propositio"; a "problem to find" is called "problema," and a "problem to prove" "theorema." In old-fashioned mathematical language, the words proposition, problem, theorem have still this "Euclidean" meaning, but this is completely changed in modern mathematical language; this justifies the introduction of new terms.

6. *Progressive reasoning* was used in various meanings by various authors, and in the old meaning of "synthesis" (see 9) by some authors. The latter usage is defensible but the term is avoided here.

7. *Regressive reasoning* was used in the old meaning of

"analysis" by some authors (compare 1, 6). The term is defensible but avoided here.

8. *Solution* is a completely clear term if taken in its purely mathematical meaning; it denotes any object satisfying the condition of a "problem to find." Thus, the solutions of the equation $x^2 - 3x + 2 = 0$ are its roots, the numbers 1 and 2. Unfortunately, the word has also other meanings which are not purely mathematical and which are used by mathematicians along with its mathematical meaning. Solution may also mean the "process of solving the problem" or the "work done in solving the problem"; we use the word in this meaning when we talk about a "difficult solution." Solution may also mean the result of the work done in solving the problem; we may use the word in this meaning when we talk about a "beautiful solution." Now, it may happen that we have to talk in the same sentence about the object satisfying the condition of the problem, about the work of obtaining it, and about the result of this work; if we yield to the temptation to call all three things "solution" the sentence cannot be too clear.

9. *Synthesis* is used by PAPPUS in a well defined meaning which would deserve to be conserved. The term is, however, regretfully avoided in the present study, for the same reasons as its counterpart "analysis" (see under 1).

Test by dimension is a well-known, quick and efficient means to check geometrical or physical formulas.

1. In order to recall the operation of the test, let us consider the frustum of a right circular cone. Let

> R be the radius of the lower base,
> r the radius of the upper base,
> h the altitude of the frustum,
> S the area of the lateral surface of the frustum.

If R, r, h are given, S is visibly determined. We find the expression

$$S = \pi(R + r)\sqrt{(R - r)^2 + h^2}$$

to which we wish to apply the test by dimension.

The dimension of a geometric quantity is easily visible. Thus, R, r, h are lengths, they are measured in centimeters if we use scientific units, their dimension is cm. The area S is measured in square centimeters, its dimension is cm^2. Now, $\pi = 3.14159 \ldots$ is a mere number; if we wish to ascribe a dimension to a purely numerical quantity it must be $cm^0 = 1$.

Each term of a sum must have the same dimension which is also the dimension of the sum. Thus, R, r, and $R + r$ have the same dimension, namely cm. The two terms $(R - r)^2$ and h^2 have the same dimension (as they must), cm^2.

The dimension of a product is the product of the dimensions of its factors, and there is a similar rule about powers. Replacing the quantities by their dimensions on both sides of the formula that we are testing, we obtain

$$cm^2 = 1 \cdot cm \cdot \sqrt{cm^2}.$$

This being visibly so, the test could not detect any error in the formula. The formula passed the test.

For other examples, see section **14**, and CAN YOU CHECK THE RESULT? 2.

2. We may apply the test by dimension to the final result of a problem or to intermediary results, to our own work or to the work of others (very suitable in tracing mistakes in examination papers), and also to formulas that we recollect and to formulas that we guess.

If you recollect the formulas $4\pi r^2$ and $4\pi r^3/3$ for the area and the volume of the sphere, but are not quite sure which is which, the test by dimension easily removes the doubt.

3. The test by dimension is even more important in physics than in geometry.

Let us consider a "simple" pendulum, that is, a small heavy body suspended by a wire whose length we regard as invariable and whose weight we regard as negligible. Let l stand for the length of the wire, g for the gravitational acceleration, and T for the period of the pendulum.

Mechanical considerations show that T depends on l and g alone. But what is the form of the dependence? We may remember or guess that

$$T = cl^m g^n$$

where c, m, n are certain numerical constants. That is, we suppose that T is proportional to certain powers, l^m, g^n, of l and g.

We look at the dimensions. As T is a time, it is measured in seconds, its dimension is *sec*. The dimension of the length l is *cm*, the dimension of the acceleration g is $cm\ sec^{-2}$, and the dimension of the numerical constant c is 1. The test by dimension yields the equation

$$sec = 1 \cdot (cm)^m\,(cm\ sec^{-2})^n$$

or

$$sec = (cm)^{m+n}\,sec^{-2n}.$$

Now, we must have the same powers of the fundamental units *cm* and *sec* on both sides, and thus we obtain

$$0 = m + n \qquad 1 = -2n$$

and hence

$$n = -\frac{1}{2} \qquad m = \frac{1}{2}\,.$$

Therefore, the formula for the period T must have the form

$$T = cl^{\frac{1}{2}}g^{-\frac{1}{2}} = c\sqrt{\frac{l}{g}}$$

The test by dimension yields much in this case but it cannot yield everything. First, it gives no information about the value of the constant c (which is, in fact, 2π). Second, it gives no information about the limits of validity; the formula is valid only for small oscillations of the pendulum and only approximately (it is exact for "infinitely small" oscillations). In spite of these limitations, there is no doubt that the consideration of the dimensions has allowed us to foresee quickly and with the most elementary means an essential part of a result whose exhaustive treatment demands much more advanced means. And this is so in many similar cases.

The future mathematician should be a clever problem-solver; but to be a clever problem-solver is not enough. In due time, he should solve significant mathematical problems; and first he should find out for which kind of problems his native gift is particularly suited.

For him, the most important part of the work is to look back at the completed solution. Surveying the course of his work and the final shape of the solution, he may find an unending variety of things to observe. He may meditate upon the difficulty of the problem and about the decisive idea; he may try to see what hampered him and what helped him finally. He may look out for simple intuitive ideas: *Can you see it at a glance?* He may compare and develop various methods: *Can you derive the result differently?* He may try to clarify his present problem by comparing it to problems formerly solved; he may try to invent new problems which he can solve on the basis of his just completed work: *Can you use the result, or the method, for some other problem?* Digesting the problems he solved as completely as he can, he may acquire well ordered knowledge, ready to use.

The future mathematician learns, as does everybody

else, by imitation and practice. He should look out for the right model to imitate. He should observe a stimulating teacher. He should compete with a capable friend. Then, what may be the most important, he should read not only current textbooks but good authors till he finds one whose ways he is naturally inclined to imitate. He should enjoy and seek what seems to him simple or instructive or beautiful. He should solve problems, choose the problems which are in his line, meditate upon their solution, and invent new problems. By these means, and by all other means, he should endeavor to make his first important discovery: he should discover his likes and his dislikes, his taste, his own line.

The intelligent problem-solver often asks himself questions similar to those contained in our list. He, perhaps, discovered questions of this sort by himself; or, having heard such a question from somebody, he discovered its proper use by himself. He is possibly not conscious at all that he repeats the same stereotyped question again and again. Or the question is his particular pet; he knows that the question is part of his mental attitude appropriate in such and such a phase of the work, and he summons up the right attitude by asking the right question.

The intelligent problem-solver may find the questions and suggestions of our list useful. He may understand quite well the explanations and examples illustrating a certain question, he may suspect the proper use of the question; but he cannot attain real understanding unless he comes across the procedure that the question tries to provoke in his own work and, by having experienced its usefulness, discovers the proper use of the question for himself.

The intelligent problem-solver should be prepared to ask all questions of the list but he should ask none unless he is prompted to do so by careful consideration of the problem at hand and by his own unprejudiced judgment. In fact, he must recognize by himself whether the present situation is sufficiently similar or not to some other situation in which he saw the question successfully applied.

The intelligent problem-solver tries first of all to understand the problem as fully and as clearly as he can. Yet understanding alone is not enough; he must concentrate upon the problem, he must desire earnestly to obtain its solution. If he cannot summon up real desire for solving the problem he would do better to leave it alone. The open secret of real success is to throw your whole personality into your problem.

The intelligent reader of a mathematical book desires two things:

First, to see that the present step of the argument is correct.

Second, to see the purpose of the present step.

The intelligent listener to a mathematical lecture has the same wishes. If he cannot see that the present step of the argument is correct and even suspects that it is, possibly, incorrect, he may protest and ask a question. If he cannot see any purpose in the present step, nor suspect any reason for it, he usually cannot even formulate a clear objection, he does not protest, he is just dismayed and bored, and loses the thread of the argument.

The intelligent teacher and the intelligent author of textbooks should bear these points in mind. To write and speak correctly is certainly necessary; but it is not sufficient. A derivation correctly presented in the book or on the blackboard may be inaccessible and uninstructive, if the purpose of the successive steps is incomprehensible, if the reader or listener cannot understand how it was humanly possible to find such an argument, if he is not able to derive any suggestion from the presentation as to how he could find such an argument by himself.

The questions and suggestions of our list may be useful to the author and to the teacher in emphasizing the purpose and the motives of his argument. Particularly useful in this respect is the question: DID WE USE ALL THE DATA? The author or the teacher may show by this question a good reason for considering the datum that has not been used heretofore. The reader or the listener can use the same question in order to understand the author's or the teacher's reason for considering such and such an element, and he may feel that, asking this question, he could have discovered this step of the argument by himself.

The traditional mathematics professor of the popular legend is absentminded. He usually appears in public with a lost umbrella in each hand. He prefers to face the blackboard and to turn his back on the class. He writes *a*, he says *b*, he means *c*; but it should be *d*. Some of his sayings are handed down from generation to generation.

"In order to solve this differential equation you look at it till a solution occurs to you."

"This principle is so perfectly general that no particular application of it is possible."

"Geometry is the art of correct reasoning on incorrect figures."

"My method to overcome a difficulty is to go round it."

"What is the difference between method and device? A method is a device which you use twice."

After all, you can learn something from this traditional mathematics professor. Let us hope that the mathematics teacher from whom you cannot learn anything will not become traditional.

Variation of the problem. An insect (as mentioned elsewhere) tries to escape through the windowpane, tries the same hopeless thing again and again, and does not try the next window which is open and through which it came into the room. A mouse may act more intelligently; caught in the trap, he tries to squeeze through between two bars, then between the next two bars, then between other bars; he varies his trials, he explores various possibilities. A man is able, or should be able, to vary his trials still more intelligently, to explore the various possibilities with more understanding, to learn by his errors and shortcomings. "Try, try again" is popular advice. It is good advice. The insect, the mouse, and the man follow it; but if one follows it with more success than the others it is because he *varies his problem* more intelligently.

1. At the end of our work, when we have obtained the solution, our conception of the problem will be fuller and more adequate than it was at the outset. Desiring to proceed from our initial conception of the problem to a more adequate, better adapted conception, we try various standpoints and we view the problem from different sides.

Success in solving the problem depends on choosing the right aspect, on attacking the fortress from its accessible side. In order to find out which aspect is the right one, which side is accessible, we try various sides and aspects, we *vary the problem.*

2. Variation of the problem is essential. This fact can be explained in various ways. Thus, from a certain point of view, progress in solving the problem appears as mobilization and organization of formerly acquired knowledge. We have to extract from our memory and to work into the problem certain elements. Now, variation of the problem helps us to extract such elements. How?

We remember things by a kind of "action by contact," called "mental association"; what we have in our mind at present tends to recall what was in contact with it at some previous occasion. (There is no space and no need to state more neatly the theory of association, or to discuss its limitations.) *Varying the problem,* we bring in new points, and so we create new contacts, new possibilities of contacting elements relevant to our problem.

3. We cannot hope to solve any worth-while problem without intense concentration. But we are easily tired by intense concentration of our attention upon the same point. In order to keep the attention alive, the object on which it is directed must unceasingly change.

If our work progresses, there is something to do, there are new points to examine, our attention is occupied, our interest is alive. But if we fail to make progress, our attention falters, our interest fades, we get tired of the problem, our thoughts begin to wander, and there is danger of losing the problem altogether. To escape from this danger we have to *set ourselves a new question* about the problem.

The new question unfolds untried possibilities of contact with our previous knowledge, it revives our hope of making useful contacts. The new question *reconquers our interest by varying the problem,* by showing some new aspect of it.

4. *Example.* Find the volume of the frustum of a pyramid with square base, being given the side of the lower base a, the side of the upper base b, and the altitude of the frustum h.

The problem may be proposed to a class familiar with the formulas for the volume of prism and pyramid. If the students do not come forward with some idea of their own, the teacher may begin with *varying the data* of the problem. We start from a frustum with $a > b$. What happens when b increases till it becomes equal to a? The frustum becomes a prism and the volume in question becomes a^2h. What happens when b decreases till it becomes equal to 0? The frustum becomes a pyramid and the volume in question becomes $a^2h/3$.

This variation of the data contributes, first of all, to the interest of the problem. Then, it may suggest using, in some way or other, the results quoted about prism and pyramid. At any rate, we have found definite properties of the final result; the final formula must be such that it reduces to a^2h for $b = a$ and to $a^2h/3$ for $b = 0$. It is an advantage to foresee properties of the result we are trying to obtain. Such properties may give valuable suggestions and, in any case, when we have found the final formula we shall be able to test it. We have thus, in advance, an answer to the question: CAN YOU CHECK THE RESULT? (See there, under 2.)

5. *Example.* Construct a trapezoid being given its four sides a, b, c, d.

Let a be the lower base and c the upper base; a and c are parallel but unequal, b and d are not parallel. If there is no other idea, we may begin with varying the data.

We start from a trapezoid with $a > c$. What happens when c decreases till it becomes equal to 0? The trapezoid degenerates into a triangle. Now a triangle is a familiar and simple figure, which we can construct from various data; there could be some advantage in introducing this triangle into the figure. We do so by drawing just one auxiliary line, a diagonal of the trapezoid (Fig. 21). Examining the triangle we find however that it is scarcely useful; we know two sides, a and d, but we should have three data.

Let us try something else. What happens when c increases till it becomes equal to a? The trapezoid becomes

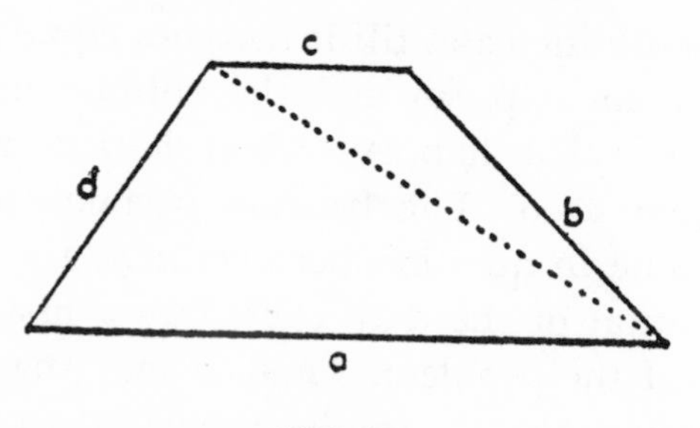

FIG. 21

a parallelogram. Could we use it? A little examination (see Fig. 22) directs our attention to the triangle which we have added to the original trapezoid when drawing the parallelogram. This triangle is easily constructed; we know three data, its three sides b, d, and $a - c$.

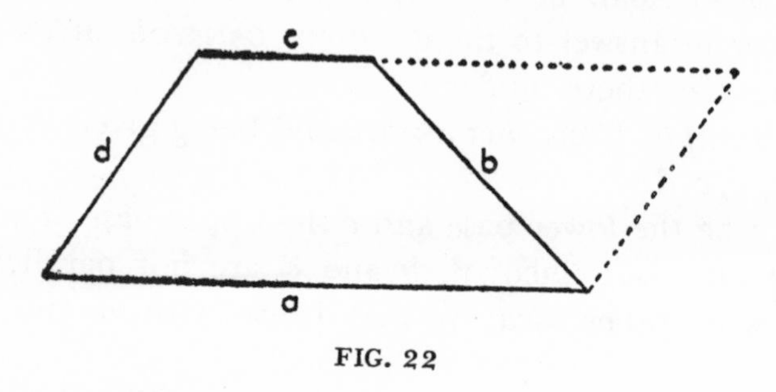

FIG. 22

Varying the original problem (construction of the trapezoid) we have been led to a more accessible auxiliary problem (construction of the triangle). Using the result of the auxiliary problem we easily solve our original problem (we have to complete the parallelogram).

Our example is typical. It is also typical that our first attempt failed. Looking back at it, we may see however that that first attempt was not so useless. There was some idea in it; in particular, it gave us an opportunity to think of the construction of a triangle as means to our end. In fact, we arrived at our second, successful trial by modifying our first, unsuccessful trial. We varied c; we first tried to decrease it, then to increase it.

6. As in the foregoing example, we often have to try various modifications of the problem. We have to vary, to restate, to transform it again and again till we succeed eventually in finding something useful. We may learn by failure; there may be some good idea in an unsuccessful trial, and we may arrive at a more successful trial by *modifying* an unsuccessful one. What we attain after various trials is very often, as in the foregoing example, a more accessible auxiliary problem.

7. There are certain modes of varying the problem which are typically useful, as going back to the DEFINITION, DECOMPOSING AND RECOMBINING, introducing AUXILIARY ELEMENTS, GENERALIZATION, SPECIALIZATION, and the use of ANALOGY.

8. What we said a while ago (under 3) about new questions which may reconquer our interest is important for the proper use of our list.

A teacher may use the list to help his students. If the student progresses, he needs no help and the teacher should not ask him any questions, but allow him to work alone which is obviously better for his independence. But the teacher should, of course, try to find some suitable question or suggestion to help him when he gets stuck. Because then there is danger that the student will get tired of his problem and drop it, or lose interest and make some stupid blunder out of sheer indifference.

We may use the list in solving our own problems. To use it properly we proceed as in the former case. When our progress is satisfactory, when new remarks emerge spontaneously, it would be simply stupid to hamper our spontaneous progress by extraneous questions. But when our progress is blocked, when nothing occurs to us, there is danger that we may get tired of our problem. Then it is time to think of some general idea that could be helpful, of some question or suggestion of the list that might be suitable. And any question is welcome that has some chance of showing a new aspect of the problem; it may reconquer our interest, it may keep us working and thinking.

What is the unknown? What is required? What do you want? What are you supposed to seek?

What are the data? What is given? What have you?

What is the condition? By what condition is the unknown linked to the data?

These questions may be used by the teacher to test the understanding of the problem; the student should be able to answer them clearly. Moreover, they direct the student's attention to the principal parts of a "problem to find," the unknown, the data, the condition. As the consideration of these parts may be necessary again and again, the questions may be often repeated in the later phases of the solution. (Examples in sections **8, 10, 18, 20**; SETTING UP EQUATIONS, 3, 4; PRACTICAL PROBLEMS, 1; PUZZLES; and elsewhere.)

The questions are of the greatest importance for the problem-solver. He checks his own understanding of the problem, he focuses his attention on this or that principal part of the problem. The solution consists essentially in linking the unknown to the data. Therefore, the problem-solver has to focus those elements again and again, asking: *What is the unknown? What are the data?*

The problem may have many unknowns, or the condition may have various parts which must be considered separately, or it may be desirable to consider some datum by itself. Therefore, we may use various modifications of our questions, as: What are the unknowns? What is the first datum? What is the second datum? What are the various parts of the condition? What is the first clause of the condition?

The principal parts of a "problem to prove" are the hypothesis and the conclusion, and the corresponding questions are: *What is the hypothesis? What is the conclusion?* We may need some variation of verbal expression or modification of these frequently useful questions as: What do you assume? What are the various parts of your assumption? (Example in section **19.**)

Why proofs? There is a traditional story about Newton: As a young student, he began the study of geometry, as was usual in his time, with the reading of the Elements of Euclid. He read the theorems, saw that they were true, and omitted the proofs. He wondered why anybody should take pains to prove things so evident. Many years later, however, he changed his opinion and praised Euclid.

The story may be authentic or not, yet the question remains: Why should we learn, or teach, proofs? What is preferable: no proof at all, or proofs for everything, or some proofs? And, if only some proofs, which proofs?

1. *Complete proofs.* For a logician of a certain sort only complete proofs exist. What intends to be a proof must leave no gaps, no loopholes, no uncertainty whatever, or else it is no proof. Can we find complete proofs according to such a high standard in everyday life, or in legal procedure, or in physical science? Scarcely. Thus, it is difficult to understand how we could acquire the idea of such a strictly complete proof.

We may say, with a little exaggeration, that humanity learned this idea from one man and one book: from Euclid and his Elements. In any case, the study of the elements of plane geometry yields still the best opportunity to acquire the idea of rigorous proof.

Let us take as an example the proof of the theorem: *In any triangle, the sum of the three angles is equal to two right angles.*[11] Fig. 23, which is an inalienable mental property of most of us, needs little explanation. There is a line through the vertex A parallel to the side BC.

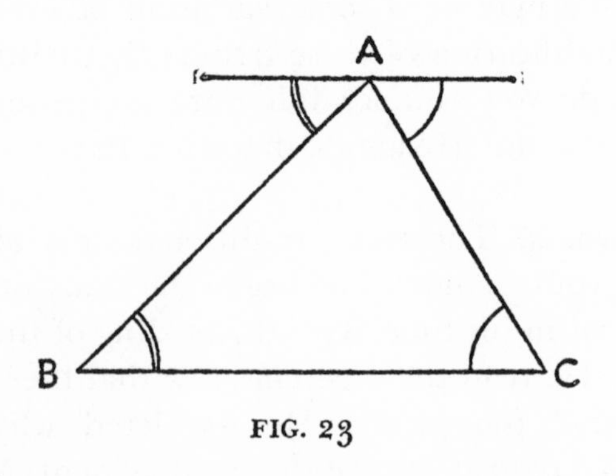

FIG. 23

The angles of the triangle at B and at C are equal to certain angles at A, as is emphasized in the figure, since alternate angles are equal in general. The three angles of the triangle are equal to three angles with a common vertex A, forming a straight angle, or two right angles; and so the theorem is proved.

If a student has gone through his mathematics classes without having really understood a few proofs like the foregoing one, he is entitled to address a scorching reproach to his school and to his teachers. In fact, we should distinguish between things of more and less importance. If the student failed to get acquainted with this or that particular geometric fact, he did not miss so much; he may have little use for such facts in later life. But if he failed to get acquainted with geometric proofs, he missed the best and simplest examples of true evidence and he missed the best opportunity to acquire the

[11] Part of Proposition 32 of Book I of Euclid's *Elements*. The following proof is not Euclid's, but was known to the Greeks.

idea of strict reasoning. Without this idea, he lacks a true standard with which to compare alleged evidence of all sorts aimed at him in modern life.

In short, if general education intends to bestow on the student the ideas of intuitive evidence and logical reasoning, it must reserve a place for geometric proofs.

2. *Logical system.* Geometry, as presented in Euclid's Elements, is not a mere collection of facts but a logical system. The axioms, definitions, and propositions are not listed in a random sequence but disposed in accomplished order. Each proposition is so placed that it can be based on the foregoing axioms, definitions, and propositions. We may regard the disposition of the propositions as Euclid's main achievement and their logical system as the main merit of the Elements.

Euclid's geometry is not only a logical system but it is the first and greatest example of such a system, which other sciences have tried, and are still trying, to imitate. Should other sciences—especially those very far from geometry, as psychology, or jurisprudence—imitate Euclid's rigid logic? This is a debatable question; but nobody can take part in the debate with competence who is not acquainted with the Euclidean system.

Now, the system of geometry is cemented with proofs. Each proposition is linked to the foregoing axioms, definitions, and propositions by a proof. Without understanding such proofs we cannot understand the very essence of the system.

In short, if general education intends to bestow on the student the idea of logical system, it must reserve a place for geometric proofs.

3. *Mnemotechnic system.* The author does not think that the ideas of intuitive evidence, strict reasoning, and logical system are superfluous for anybody. There may be cases, however, in which the study of these ideas is not

considered absolutely necessary, owing to lack of time, or for other reasons. Yet even in such cases proofs may be desirable.

Proofs yield evidence; in so doing, they hold together the logical system; and they help us to remember the various items held together. Take the example discussed above, in connection with Fig. 23. This figure renders evident the fact that the sum of the angles in a triangle equals 180°. The figure connects this fact with the other fact that alternate angles are equal. Connected facts however are more interesting and are better retained than isolated facts. Thus, our figure fixes the two connected geometric propositions in our mind and, finally, the figure and the propositions may become our inalienable mental property.

Now we come to the case in which the acquisition of general ideas is not regarded as necessary, only that of certain facts is desired. Even in such a case, the facts must be presented in some connection and in some sort of system, since isolated items are laboriously acquired and easily forgotten. Any sort of connection that unites the facts simply, naturally, and durably, is welcome here. The system need not be founded on logic, it must only be designed to aid the memory effectively; it must be what is called a *mnemotechnic* system. Yet even from the point of view of a purely mnemotechnic system, proofs may be useful, especially simple proofs. For instance, the student must learn the fact about the sum of the angles in the triangle and that other fact about the alternate angles. Can any device to retain these facts be simpler, more natural or more effective than Fig. 23?

In short, even when no special importance is attached to general logical ideas proofs may be useful as a mnemotechnic device.

4. *The cookbook system.* We have discussed the advantages of proofs but we certainly did not advocate that all

proofs should be given "in extenso." On the contrary, there are cases in which it is scarcely possible to do so; an important case is the teaching of the differential and integral calculus to students of engineering.

If the calculus is presented according to modern standards of rigor, it demands proofs of a certain degree of difficulty and subtlety ("epsilon-proofs"). But engineers study the calculus in view of its application and have neither enough time nor enough training or interest to struggle through long proofs or to appreciate subtleties. Thus, there is a strong temptation to cut out all the proofs. Doing so, however, we reduce the calculus to the level of the cookbook.

The cookbook gives a detailed description of ingredients and procedures but no proofs for its prescriptions or reasons for its recipes; the proof of the pudding is in the eating. The cookbook may serve its purpose perfectly. In fact, it need not have any sort of logical or mnemotechnic system since recipes are written or printed and not retained in memory.

Yet the author of a textbook of calculus, or a college instructor, can hardly serve his purpose if he follows the system of the cookbook too closely. If he teaches procedures without proofs, the unmotivated procedures are not understood. If he gives rules without reasons, the unconnected rules are quickly forgotten. Mathematics cannot be tested in exactly the same manner as a pudding; if all sorts of reasoning are debarred, a course of calculus may easily become an incoherent inventory of indigestible information.

5. *Incomplete proofs.* The best way of handling the dilemma between too heavy proofs and the level of the cookbook may be to make reasonable use of incomplete proofs.

For a strict logician, an incomplete proof is no proof at all. And, certainly, incomplete proofs ought to be carefully distinguished from complete proofs; to confuse one with the other is bad, to sell one for the other is worse. It is painful when the author of a textbook presents an incomplete proof ambiguously, with visible hesitation between shame and the pretension that the proof is complete. But incomplete proofs may be useful when they are used in their proper place and in good taste. Their purpose is not to replace complete proofs, which they never could, but to lend interest and coherence to the presentation.

Example 1. *An algebraic equation of degree* n *has exactly* n *roots.* This proposition, called the Fundamental Theorem of Algebra by Gauss, must often be presented to students who are quite unprepared for understanding its proof. They know however that an *equation of the first degree has one root,* and one of the *second degree two roots.* Moreover the difficult proposition has a part that can be easily shown: *no equation of degree* n *has more than* n *different roots.* Do the facts mentioned constitute a complete proof for the Fundamental Theorem? By no means. They are sufficient however to lend it a certain interest and plausibility—and to fix it in the minds of the students, which is the main thing.

Example 2. *The sum of any two of the plane angles formed by the edges of a trihedral angle is greater than the third.* Obviously, the theorem amounts to affirming that *in a spherical triangle the sum of any two sides is greater than the third.* Having observed this, we naturally think of the analogy of the spherical triangle with the rectilinear triangle. Do these remarks constitute a proof? By no means; but they help us to understand and to remember the proposed theorem.

Our first example has historical interest. For about 250 years, the mathematicians believed the Fundamental Theorem without complete proof—in fact, without much more basis than what was mentioned above. Our second example points to ANALOGY as an important source of conjectures. In mathematics, as in the natural and physical sciences, discovery often starts from observation, analogy, and induction. These means, tastefully used in framing a plausible heuristic argument, appeal particularly to the physicist and the engineer. (See also INDUCTION AND MATHEMATICAL INDUCTION, 1, 2, 3.)

The role and interest of incomplete proofs is explained to a certain extent by our study of the process of the solution. Some experience in solving problems shows that the first idea of a proof is very frequently incomplete. The most essential remark, the main connection, the germ of the proof may be there, but details must be provided afterwards and are often troublesome. Some authors, but not many, have the gift of presenting just the germ of the proof, the main idea in its simplest form, and indicating the nature of the remaining details. Such a proof, although incomplete, may be much more instructive than a proof presented with complete details.

In short, incomplete proofs may be used as a sort of mnemotechnic device (but, of course, not as substitutes for complete proofs) when the aim is tolerable coherence of presentation and not strictly logical consistency.

It is very dangerous to advocate incomplete proofs. Possible abuse, however, may be kept within bounds by a few rules. First, if a proof is incomplete, it must be indicated as such, somewhere and somehow. Second, an author or a teacher is not entitled to present an incomplete proof for a theorem unless he knows very well a complete proof for it himself.

And it may be confessed that to present an incomplete proof in good taste is not easy at all.

Wisdom of proverbs. Solving problems is a fundamental human activity. In fact, the greater part of our conscious thinking is concerned with problems. When we do not indulge in mere musing or daydreaming, our thoughts are directed toward some end; we seek means, we seek to solve a problem.

Some people are more and others less successful in attaining their ends and solving their problems. Such differences are noticed, discussed, and commented upon, and certain proverbs seem to have preserved the quintessence of such comments. At any rate, there are a good many proverbs which characterize strikingly the typical procedures followed in solving problems, the points of common sense involved, the usual tricks, and the usual errors. There are many shrewd and some subtle remarks in proverbs but, obviously, there is no scientific system free of inconsistencies and obscurities in them. On the contrary, many a proverb can be matched with another proverb giving exactly opposite advice, and there is a great latitude of interpretation. It would be foolish to regard proverbs as an authoritative source of universally applicable wisdom but it would be a pity to disregard the graphic description of heuristic procedures provided by proverbs.

It could be an interesting task to collect and group proverbs about planning, seeking means, and choosing between lines of action, in short, proverbs about solving problems. Of the space needed for such a task only a small fraction is available here; the best we can do is to quote a few proverbs illustrating the main phases of the solution emphasized in our list, and discussed in sections 6 to 14 and elsewhere. The proverbs quoted will be printed in italics.

1. The very first thing we must do for our problem is to understand it: *Who understands ill, answers ill.* We must see clearly the end we have to attain: *Think on the end before you begin.* This is an old piece of advice; "respice finem" is the saying in Latin. Unfortunately, not everybody heeds such good advice, and people often start

speculating, talking, and even acting fussily without having properly understood the aim for which they should work. *A fool looks to the beginning, a wise man regards the end.* If the end is not clear in our mind, we may easily stray from the problem and drop it. *A wise man begins in the end, a fool ends in the beginning.*

Yet it is not enough to understand the problem, we must also desire its solution. We have no chance to solve a difficult problem without a strong desire to solve it, but with such desire there is a chance. *Where there is a will there is a way.*

2. Devising a plan, conceiving the idea of an appropriate action, is the main achievement in the solution of a problem.

A good idea is a piece of good fortune, an inspiration, a gift of the gods, and we have to deserve it: *Diligence is the mother of good luck. Perseverance kills the game. An oak is not felled at one stroke. If at first you don't succeed, try, try again.* It is not enough however to try repeatedly, we must try different means, vary our trials. *Try all the keys in the bunch. Arrows are made of all sorts of wood.* We must adapt our trials to the circumstances. *As the wind blows you must set your sail. Cut your coat according to the cloth. We must do as we may if we can't do as we would.* If we have failed, we must try something else. *A wise man changes his mind, a fool never does.* We should even be prepared from the outset for a possible failure of our scheme and have another one in reserve. *Have two strings to your bow.* We may, of course, overdo this sort of changing from one scheme to another and lose time. Then we may hear the ironical comment: *Do and undo, the day is long enough.* We are likely to blunder less if we do not lose sight of our aim. *The end of fishing is not angling but catching.*

We work hard to extract something helpful from our memory, yet, quite often, when an idea that could be helpful presents itself, we do not appreciate it, for it is so inconspicuous. The expert has, perhaps, no more ideas than the inexperienced, but appreciates more what he has and uses it better. *A wise man will make more opportunities than he finds. A wise man will make tools of what comes to hand. A wise man turns chance into good fortune.* Or, possibly, the advantage of the expert is that he is continually on the lookout for opportunities. *Have an eye to the main chance.*

3. We should start carrying out our plan at the right moment, when it is ripe, but not before. We should not start rashly. *Look before you leap. Try before you trust. A wise delay makes the road safe.* On the other hand, we should not hesitate too long. *If you will sail without danger you must never put to sea. Do the likeliest and hope the best. Use the means and God will give the blessing.*

We must use our judgment to determine the right moment. And here is a timely warning that points out the most common fallacy, the most usual failure of our judgment: *We soon believe what we desire.*

Our plan gives usually but a general outline. We have to convince ourselves that the details fit into the outline, and so we have to examine carefully each detail, one after the other. *Step after step the ladder is ascended. Little by little as the cat ate the flickle. Do it by degrees.*

In carrying out our plan we must be careful to arrange its steps in the proper order, which is frequently just the reverse of the order of invention. *What a fool does at last, a wise man does at first.*

4. Looking back at the completed solution is an important and instructive phase of the work. *He thinks not well that thinks not again. Second thoughts are best.*

Reexamining the solution, we may discover an additional confirmation of the result. Yet it must be pointed out to the beginner that such an additional confirmation is valuable, that two proofs are better than one. *It is safe riding at two anchors.*

5. We have by no means exhausted the comments of proverbs on the solution of problems. Yet many other proverbs which could be quoted would scarcely furnish new themes, only variations on the themes already mentioned. Certain more systematic and more sophisticated aspects of the process of solution are hardly within the scope of the Wisdom of Proverbs.

In describing the more systematic aspects of the solution, the author tried now and then to imitate the peculiar turn of proverbs, which is not easy. Here follow a few "synthetic" proverbs which describe somewhat more sophisticated attitudes.

The end suggests the means.

Your five best friends are What, Why, Where, When, and How. You ask What, you ask Why, you ask Where, When, and How—and ask nobody else when you need advice.

Do not believe anything but doubt only what is worth doubting.

Look around when you have got your first mushroom or made your first discovery; they grow in clusters.

Working backwards. If we wish to understand human behavior we should compare it with animal behavior. Animals also "have problems" and "solve problems." Experimental psychology has made essential progress in the last decades in exploring the "problem-solving" activities of various animals. We cannot discuss here these investigations but we shall describe sketchily just one simple and instructive experiment and our description will serve as a sort of comment upon the method of analysis, or method of "working backwards." This method, by the way, is discussed also elsewhere in the present book, under the name of PAPPUS to whom we owe an important description of the method.

1. Let us try to find an answer to the following tricky question: *How can you bring up from the river exactly six quarts of water when you have only two containers, a four quart pail and a nine quart pail, to measure with?*

Let us visualize clearly the given tools we have to work with, the two containers. (*What is given?*) We imagine two cylindrical containers having equal bases whose altitudes are as 9 to 4, see Fig. 24. *If* along the lateral sur-

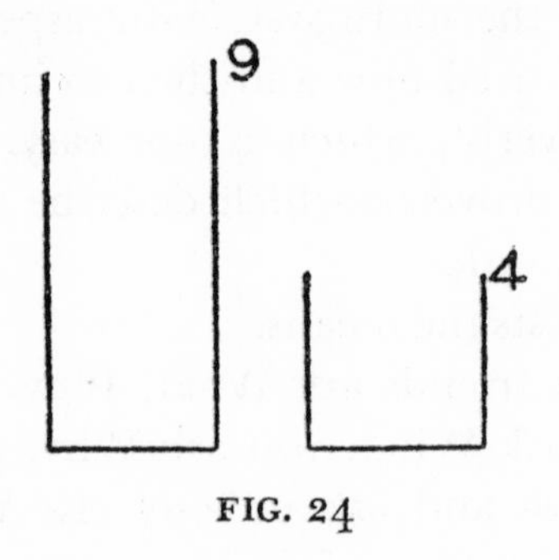

FIG. 24

face of each container there were a scale of equally spaced horizontal lines from which we could tell the height of the waterline, our problem would be easy. Yet there is no such scale and so we are still far from the solution.

We do not know yet how to measure exactly 6 quarts; but could we measure something else? (*If you cannot solve the proposed problem try to solve first some related problem. Could you derive something useful from the data?*) Let us do something, let us play around a little. We could fill the larger container to full capacity and empty so much as we can into the smaller container; then we could get 5 quarts. Could we also get 6 quarts? Here are again the two empty containers. We could also . . .

We are working now as most people do when confronted with this puzzle. We start with the two empty containers, we try this and that, we empty and fill, and when we do not succeed, we start again, trying something else. We are *working forwards,* from the given initial situation to the desired final situation, from the data to the unknown. We may succeed, after many trials, accidentally.

2. But exceptionally able people, or people who had the chance to learn in their mathematics classes something more than mere routine operations, do not spend too much time in such trials but turn around, and start working backwards.

What are we required to do? (*What is the unknown?*) Let us visualize the final situation we aim at as clearly as possible. Let us imagine that we have here, before us,

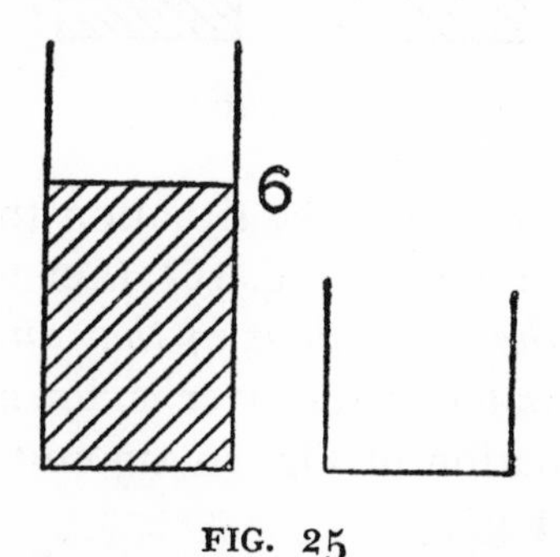

FIG. 25

the larger container with exactly 6 quarts in it and the smaller container empty as in Fig. 25. (Let us *start from what is required* and *assume what is sought as already found,* says Pappus.)

From what foregoing situation could we obtain the desired final situation shown in Fig. 25? (Let us *inquire from what antecedent the desired result could be derived,* says Pappus.) We could, of course, fill the larger container to full capacity, that is, to 9 quarts. But then we should be able to pour out exactly three quarts. To do that . . . we must have just one quart in the smaller container! That's the idea. See Fig. 26.

(The step that we have just completed is not easy at all. Few persons are able to take it without much foregoing hesitation. In fact, recognizing the significance of this step, we foresee an outline of the following solution.)

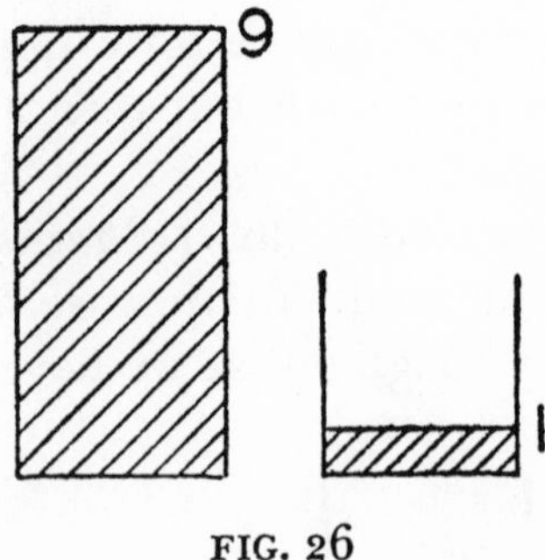

FIG. 26

But how can we reach the situation that we have just found and illustrated by Fig. 26? (Let us *inquire again what could be the antecedent of that antecedent.*) Since the amount of water in the river is, for our purpose, unlimited, the situation of Fig. 26 amounts to the same as the next one in Fig. 27

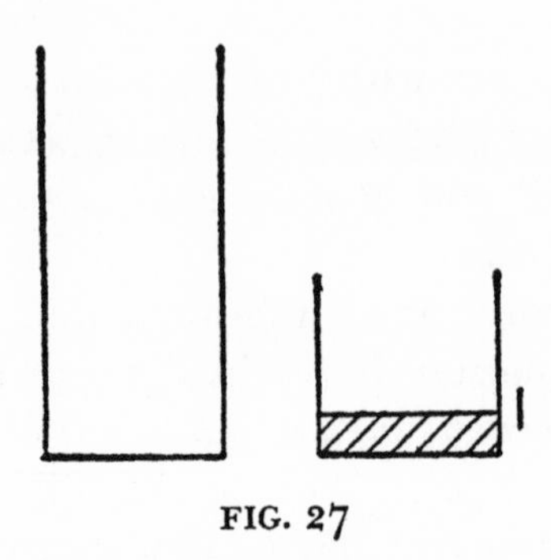

FIG. 27

or the following in Fig. 28.

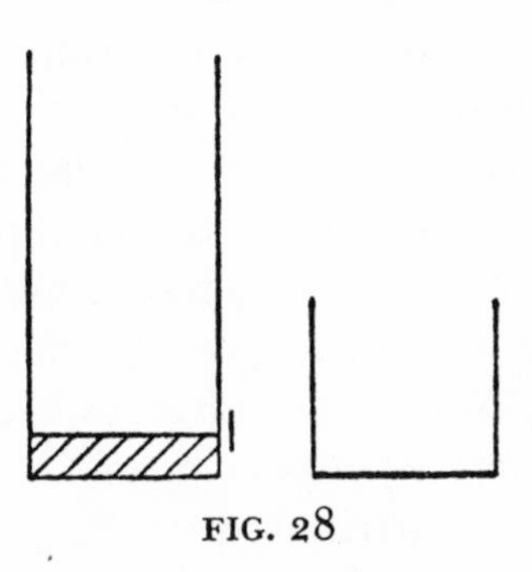

FIG. 28

It is easy to recognize that if any one of the situations in Figs. 26, 27, 28 is obtained, any other can be obtained just as well, but it is not so easy to hit upon Fig. 28, unless we *have seen it before,* encountered it accidentally in one of our initial trials. Playing around with the two containers, we may have done something similar and remember now, in the right moment, that the situation of Fig. 28 can arise as suggested by Fig. 29: We fill the large

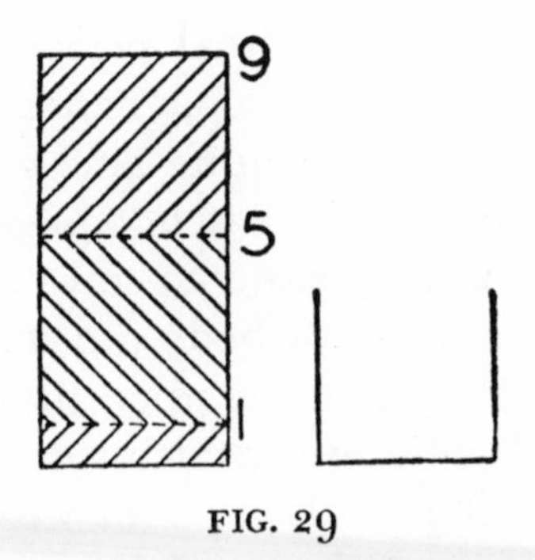

FIG. 29

container to full capacity, and pour from it four quarts into the smaller container and then into the river, twice in succession. We *came eventually upon something already known* (these are Pappus's words) and following the method of analysis, *working backwards,* we have discovered the appropriate sequence of operations.

It is true, we have discovered the appropriate sequence in retrogressive order but all that is left to do is to *reverse the process* and *start from the point which we reached last of all in the analysis* (as Pappus says). First, we perform the operations suggested by Fig. 29 and obtain Fig. 28; then we pass to Fig. 27, then to Fig. 26, and finally to Fig. 25. *Retracing our steps, we finally succeed in deriving what was required.*

3. Greek tradition attributed to Plato the discovery of the method of analysis. The tradition may not be quite reliable but, at any rate, if the method was not invented by Plato, some Greek scholar found it necessary to attribute its invention to a philosophical genius.

There is certainly something in the method that is not superficial. There is a certain psychological difficulty in turning around, in going away from the goal, in working backwards, in not following the direct path to the desired end. When we discover the sequence of appropriate operations, our mind has to proceed in an order which is exactly the reverse of the actual performance. There is some sort of psychological repugnance to this reverse order which may prevent a quite able student from understanding the method if it is not presented carefully.

Yet it does not take a genius to solve a concrete problem working backwards; anybody can do it with a little common sense. We concentrate upon the desired end, we visualize the final position in which we would like to be. From what foregoing position could we get there? It is natural to ask this question, and in so asking we work backwards. Quite primitive problems may lead naturally to working backwards; see PAPPUS, 4.

Working backwards is a common-sense procedure within the reach of everybody and we can hardly doubt that it was practiced by mathematicians and nonmathematicians before Plato. What some Greek scholar may have regarded as an achievement worthy of the genius of Plato is to state the procedure in general terms and to stamp it as an operation typically useful in solving mathematical and nonmathematical problems.

4. And now, we turn to the psychological experiment—if the transition from Plato to dogs, hens, and chimpanzees is not too abrupt. A fence forms three sides of a rectangle but leaves open the fourth side as shown in Fig. 30. We place a dog on one side of the fence, at the

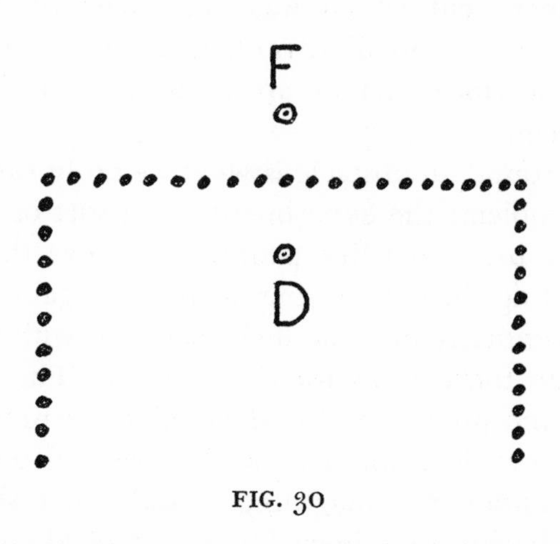

FIG. 30

point *D,* and some food on the other side, at the point *F.* The problem is fairly easy for the dog. He may first strike a posture as if to spring directly at the food but then he quickly turns about, dashes off around the end of the fence and, running without hesitation, reaches the food in a smooth curve. Sometimes, however, especially when the points *D* and *F* are close to each other, the solution is not so smooth; the dog may lose some time in barking, scratching, or jumping against the fence before he "conceives the bright idea" (as we would say) of going around.

It is interesting to compare the behavior of various animals put into the place of the dog. The problem is very easy for a chimpanzee or a four-year-old child (for whom a toy may be a more attractive lure than food). The problem, however, turns out to be surprisingly difficult for a hen who runs back and forth excitedly on her side of the fence and may spend considerable time before getting at the food if she gets there at all. But she may succeed, after much running, accidentally.

5. We should not build a big theory upon just one simple experiment which was only sketchily reported. Yet there can be no disadvantage in noticing obvious analogies provided that we are prepared to recheck and revalue them.

Going around an obstacle is what we do in solving any kind of problem; the experiment has a sort of symbolic value. The hen acted like people who solve their problem muddling through, trying again and again, and succeeding eventually by some lucky accident without much insight into the reasons for their success. The dog who scratched and jumped and barked before turning around solved his problem about as well as we did ours about the two containers. Imagining a scale that shows the waterline in our containers was a sort of almost useless scratching, showing only that what we seek lies deeper under the surface. We also tried to work forwards first, and came to the idea of turning round afterwards. The dog who, after brief inspection of the situation, turned round and dashed off gives, rightly or wrongly, the impression of superior insight.

No, we should not even blame the hen for her clumsiness. There is a certain difficulty in turning round, in going away from the goal, in proceeding without looking continually at the aim, in not following the direct path to the desired end. There is an obvious analogy between her difficulties and our difficulties.

PART IV. PROBLEMS, HINTS, SOLUTIONS

This last part offers the reader additional opportunity for practice.

The problems require no more preliminary knowledge than the reader could have acquired from a good high-school curriculum. Yet they are not too easy and not mere routine problems; some of them demand originality and ingenuity.[12]

The hints offer indications leading to the result, mostly by quoting an appropriate sentence from the list; to a very attentive reader ready to pick up suggestions they may reveal the key idea of the solution.

The solutions bring not only the answer but also the procedure leading to the answer, although, of course, the reader has to supply some of the details. Some solutions try to open up some further outlook by a few words placed at the end.

The reader who has earnestly tried to solve the problem has the best chance to profit by the hint and the solution. If he obtains the result by his own means, he may learn something by comparing his method with the method given in print. If, after a serious effort, he is inclined to give up, the hint may supply him with the

[12] Except Problem 1 (widely known, but too amusing to miss) all the problems are taken from the Stanford University Competitive Examinations in Mathematics (there are a few minor changes). Some of the problems were formerly published in *The American Mathematical Monthly* and/or *The California Mathematics Council Bulletin*. In the latter periodical also some solutions were published by the author; they appear appropriately rearranged in the sequel.

missing idea. If even the hint does not help, he may look at the solution, try to isolate the key idea, put the book aside, and then try to work out the solution.

PROBLEMS

1. A bear, starting from the point P, walked one mile due south. Then he changed direction and walked one mile due east. Then he turned again to the left and walked one mile due north, and arrived exactly at the point P he started from. What was the color of the bear?

2. Bob wants a piece of land, exactly level, which has four boundary lines. Two boundary lines run exactly north-south, the two others exactly east-west, and each boundary line measures exactly 100 feet. Can Bob buy such a piece of land in the U.S.?

3. Bob has 10 pockets and 44 silver dollars. He wants to put his dollars into his pockets so distributed that each pocket contains a different number of dollars. Can he do so?

4. To number the pages of a bulky volume, the printer used 2989 digits. How many pages has the volume?

5. Among Grandfather's papers a bill was found:

72 turkeys $_67.9_

The first and last digit of the number that obviously represented the total price of those fowls are replaced here by blanks, for they have faded and are now illegible.

What are the two faded digits and what was the price of one turkey?

6. Given a regular hexagon and a point in its plane. Draw a straight line through the given point that divides the given hexagon into two parts of equal area.

7. Given a square. Find the locus of the points from

which the square is seen under an angle (a) of $90°$ (b) of $45°$. (Let P be a point outside the square, but in the same plane. The smallest angle with vertex P containing the square is the "angle under which the square is seen" from P.) Sketch clearly both loci and give a full description.

8. Call "axis" of a solid a straight line joining two points of the surface of the solid and such that the solid, rotated about this line through an angle which is greater than $0°$ and less than $360°$ coincides with itself.

Find the axes of a cube. Describe clearly the location of the axes, find the angle of rotation associated with each. Assuming that the edge of the cube is of unit length, compute the arithmetic mean of the lengths of the axes.

9. In a tetrahedron (which is not necessarily regular) two opposite edges have the same length a and they are perpendicular to each other. Moreover they are each perpendicular to a line of length b which joins their midpoints. Express the volume of the tetrahedron in terms of a and b, and prove your answer.

10. The vertex of a pyramid opposite the base is called the *apex*. (a) Let us call a pyramid "isosceles" if its apex is at the same distance from all *vertices* of the base. Adopting this definition, prove that the base of an isosceles pyramid is *inscribed* in a circle the center of which is the foot of the pyramid's altitude.

(b) Now let us call a pyramid "isosceles" if its apex is at the same (perpendicular) distance from all sides of the base. Adopting this definition (different from the foregoing) prove that the base of an isosceles pyramid is *circumscribed* about a circle the center of which is the foot of the pyramid's altitude.

11. Find x, y, u, and v, satisfying the system of four equations

$$\begin{aligned} x + 7y + 3v + 5u &= 16 \\ 8x + 4y + 6v + 2u &= -16 \\ 2x + 6y + 4v + 8u &= 16 \\ 5x + 3y + 7v + u &= -16 \end{aligned}$$

(This may look long and boring: look for a short cut.)

12. Bob, Peter, and Paul travel together. Peter and Paul are good hikers; each walk p miles per hour. Bob has a bad foot and drives a small car in which two people can ride, but not three; the car covers c miles per hour. The three friends adopted the following scheme: They start together, Paul rides in the car with Bob, Peter walks. After a while, Bob drops Paul, who walks on; Bob returns to pick up Peter, and then Bob and Peter ride in the car till they overtake Paul. At this point they change: Paul rides and Peter walks just as they started and the whole procedure is repeated as often as necessary.

(a) How much progress (how many miles) does the company make per hour?

(b) Through which fraction of the travel time does the car carry just one man?

(c) Check the extreme cases $p = 0$ and $p = c$.

13. Three numbers are in arithmetic progression, three other numbers in geometric progression. Adding the corresponding terms of these two progressions successively, we obtain

$$85, \quad 76, \quad \text{and} \quad 84$$

respectively, and, adding all three terms of the arithmetic progression, we obtain 126. Find the terms of both progressions.

14. Determine m so that the equation in x

$$x^4 - (3m + 2)x^2 + m^2 = 0$$

has four real roots in arithmetic progression.

15. The length of the perimeter of a right triangle is 60 inches and the length of the altitude perpendicular to the hypotenuse is 12 inches. Find the sides.

16. From the peak of a mountain you see two points, A and B, in the plain. The lines of vision, directed to these points, include the angle γ. The inclination of the first line of vision to a horizontal plane is α, that of the second line β. It is known that the points A and B are on the same level and that the distance between them is c.

Express the elevation x of the peak above the common level of A and B in terms of the angles α, β, γ, and the distance c.

17. Observe that the value of

$$\frac{1}{2!} + \frac{2}{3!} + \frac{3}{4!} + \cdots + \frac{n}{(n+1)!}$$

is $1/2$, $5/6$, $23/24$ for $n = 1,2,3$, respectively, guess the general law (by observing more values if necessary) and prove your guess.

18. Consider the table

$$\begin{array}{rcr} 1 & = & 1 \\ 3+5 & = & 8 \\ 7+9+11 & = & 27 \\ 13+15+17+19 & = & 64 \\ 21+23+25+27+29 & = & 125 \end{array}$$

Guess the general law suggested by these examples, express it in suitable mathematical notation, and prove it.

19. The side of a regular hexagon is of length n (n is an integer). By equidistant parallels to its sides the hexagon is divided into T equilateral triangles each of which has sides of length 1. Let V denote the number of vertices appearing in this division, and L the number of boundary lines of length 1. (A boundary line belongs to one or two triangles, a vertex to two or more triangles.) When

$n = 1$, which is the simplest case, $T = 6$, $V = 7$, $L = 12$. Consider the general case and express T, V, and L in terms of n. (Guessing is good, proving is better.)

20. In how many ways can you change one dollar? (The "way of changing" is determined if it is known how many coins of each kind—cents, nickels, dimes, quarters, half dollars—are used.)

HINTS

1. *What is the unknown?* The color of a bear—but how could we find the color of a bear from mathematical data? *What is given?* A geometrical situation—but it seems self-contradictory: how could the bear, after walking three miles in the manner described, return to his *starting* point?

2. *Do you know a related problem?*

3. If Bob had very many dollars, he would have obviously no difficulty in filling each of his pockets differently. *Could you restate the problem?* What is the minimum number of dollars that can be put in 10 pockets so that no two different pockets contain the same amount?

4. *Here is a problem related to yours:* If the book has exactly 9 numbered pages, how many digits uses the printer? (9, of course.) Here is another *problem related to yours:* If the book has exactly 99 numbered pages, how many digits does the printer use?

5. *Could you restate the problem?* What can the two faded digits be if the total price, expressed in cents, is divisible by 72?

6. *Could you imagine a more accessible related problem? A more general problem? An analogous problem?* (GENERALIZATION, 2.)

7. *Do you know a related problem?* The locus of the points from which a given segment of a straight line is

seen under a given angle consists of two circular arcs, ending in the extreme points of the segment, and symmetric to each other with respect to the segment.

8. I assume that the reader is familiar with the shape of the cube and has found certain axes just by inspection —but are they *all* the axes? *Can you prove* that your list of axes is exhaustive? Has your list a clear principle of classification?

9. *Look at the unknown!* The unknown is the volume of a tetrahedron—yes, I know, the volume of any pyramid can be computed when the base and the height are given (product of both, divided by 3) but in the present case neither the base nor the height is given. *Could you imagine a more accessible related problem?* (Don't you see a more accessible tetrahedron which is an aliquot part of the given one?)

10. *Do you know a related theorem?* Do you know a related . . . simpler . . . *analogous theorem?* Yes: the foot of the altitude is the mid-point of the base in an isosceles triangle. Here is a *theorem related to yours and proved before. Could you use . . . its method?* The theorem on the isosceles triangle is proved from congruent right triangles of which the altitude is a common side.

11. It is assumed that the reader is somewhat familiar with systems of linear equations. To solve such a system, we have to combine its equations in some way—look out for relations between the equations which could indicate a particularly advantageous combination.

12. *Separate the various parts of the condition. Can you write them down?* Between the start and the point where the three friends meet again there are three different phases:

(1) Bob rides with Paul
(2) Bob rides alone
(3) Bob rides with Peter.

Call t_1, t_2, and t_3 the durations of these phases, respectively. How could you split the condition into appropriate parts?

13. *Separate the various parts of the condition. Can you write them down?* Let

$$a - d, \qquad a, \qquad a + d$$

be the terms of the arithmetic progression, and

$$bg^{-1}, \qquad b, \qquad bg$$

be the terms of the geometric progression.

14. *What is the condition?* The four roots must form an arithmetic progression. Yet the equation has a particular feature: it contains only even powers of the unknown x. Therefore, if a is a root, $-a$ is also a root.

15. *Separate the various parts of the condition. Can you write them down?* We may distinguish three parts in the condition, concerning

(1) perimeter
(2) right triangle
(3) height to hypotenuse.

16. *Separate the various parts of the condition. Can you write them down?* Let a and b stand for the lengths of the (unknown) lines of vision, α and β for their inclinations to the horizontal plane, respectively. We may distinguish three parts in the condition, concerning

(1) the inclination of a
(2) the inclination of b
(3) the triangle with sides a, b, and c.

17. Do you *recognize* the denominators 2, 6, 24? *Do you know a related problem? An analogous problem?* (INDUCTION AND MATHEMATICAL INDUCTION.)

18. Discovery by induction needs observation. Observe the right-hand sides, the initial terms of the left-hand sides, and the final terms. What is the general law?

19. *Draw a figure.* Its observation may help you to discover the law inductively, or it may lead you to relations between T, V, L, and n.

20. *What is the unknown?* What are we supposed to seek? Even the aim of the problem may need some clarification. *Could you imagine a more accessible related problem? A more general problem? An analogous problem?* Here is a *very* simple analogous problem: In how many ways can you pay one cent? (There is just one way.) Here is a more general problem: In how many ways can you pay the amount of n cents using these five kinds of coins: cents, nickels, dimes, quarters, and half dollars. We are especially concerned with the particular case $n = 100$.

In the simplest particular cases, for small n, we can figure out the answer without any high-brow method, just by trying, by inspection. Here is a short table (which the reader should check).

n	4	5	9	10	14	15	19	20	24	25
E_n	1	2	2	4	4	6	6	9	9	13

The first line lists the amounts to be paid, generally called n. The second line lists the corresponding numbers of "ways of paying," generally called E_n. (Why I have chosen this notation is a secret of mine which I am not willing to give away at this stage.)

We are especially concerned with E_{100}, but there is little hope that we can compute E_{100} without some clear method. In fact the present problem requires a little more from the reader than the foregoing ones; he should *create* a little *theory*.

Our question is general (to compute E_n for general n),

but it is "isolated." *Could you imagine a more accessible related problem? An analogous problem?* Here is a *very* simple analogous problem: Find A_n, the number of ways to pay the amount of n cents, using only cents. ($A_n = 1$.)

SOLUTIONS

1. You think that the bear was white and the point P is the North Pole? *Can you prove that this is correct?* As it was more or less understood, we idealize the question. We regard the globe as exactly spherical and the bear as a moving material point. This point, moving due south or due north, describes an arc of a *meridian* and it describes an arc of a *parallel* circle (parallel to the equator) when it moves due east. We have to distinguish two cases.

(1) If the bear returns to the point P along a meridian *different* from the one along which he left P, P is necessarily the North Pole. In fact the only other point of the globe in which two meridians meet is the South Pole, but the bear could leave this pole only in moving northward.

(2) The bear could return to the point P along the same meridian he left P if, when walking one mile due east, he describes a parallel circle exactly n times, where n may be 1, 2, 3 . . . In this case P is not the North Pole, but a point on a parallel circle very close to the South Pole (the perimeter of which, expressed in miles, is slightly inferior to $2\pi + 1/n$).

2. We represent the globe as in the solution of Problem 1. The land that Bob wants is bounded by two meridians and two parallel circles. Imagine two fixed meridians, and a parallel circle moving *away* from the equator: the arc on the moving parallel intercepted by the two fixed meridians is steadily shortened. The center of the land that Bob wants should be on the equator: he can *not* get it in the U.S.

3. The least possible number of dollars in a pocket is obviously 0. The next greater number is at least 1, the next greater at least 2 . . . and the number in the last (tenth) pocket is at least 9. Therefore, the number of dollars required is at least

$$0 + 1 + 2 + 3 + \dots + 9 = 45$$

Bob cannot make it: he has only 44 dollars.

4. A volume of 999 numbered pages needs

$$9 + 2 \times 90 + 3 \times 900 = 2889$$

digits. If the bulky volume in question has x pages

$$2889 + 4(x - 999) = 2989$$
$$x = 1024$$

This problem may teach us that a preliminary estimate of the unknown may be useful (or even necessary, as in the present case).

5. If _679_ is divisible by 72, it is divisible both by 8 and by 9. If it is divisible by 8, the number 79_ must be divisible by 8 (since 1000 is divisible by 8) and so 79_ must be 792: the last faded digit is 2. If _6792 is divisible by 9, the sum of its digits must be divisible by 9 (the rule about "casting out nines") and so the first faded digit must be 3. The price of one turkey was (in grandfather's time) $367.92 ÷ 72 = $5.11.

6. "*A point and a figure with a center of symmetry* (in the same plane) are given in position. Find a straight line that passes through the given point and bisects the area of the given figure." The required line passes, of course, through the center of symmetry. See INVENTOR'S PARADOX.

7. In any position the two sides of the angle must pass through two vertices of the square. As long as they pass through the same pair of vertices, the angle's vertex

moves along the same arc of circle (by the theorem underlying the hint). Hence each of the two loci required consists of several arcs of circle: of 4 semicircles in the case (a) and of 8 quarter circles in the case (b); see Fig. 31.

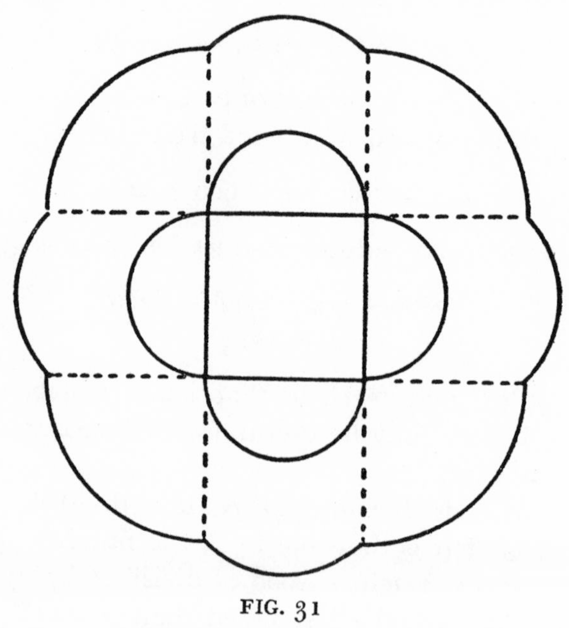

FIG. 31

8. The axis pierces the surface of the cube in some point which is either a vertex of the cube or lies on an edge or in the interior of a face. If the axis passes through a point of an edge (but not through one of its endpoints) this point must be the midpoint: otherwise the edge could not coincide with itself after the rotation. Similarly, an axis piercing the interior of a face must pass through its center. Any axis must, of course, pass through the center of the cube. And so there are three kinds of axes:

(1) 4 axes, each through two opposite vertices; angles 120°, 240°

(2) 6 axes, each through the mid-points of two opposite edges; angle 180°

(3) 3 axes, each through the center of two opposite faces; angles 90°, 180°, 270°.

For the length of an axis of the first kind see section 12; the others are still easier to compute. The desired average is

$$\frac{4\sqrt{3} + 6\sqrt{2} + 3}{13} = 1.416.$$

(This problem may be useful in preparing the reader for the study of crystallography. For the reader sufficiently advanced in the integral calculus it may be observed that the average computed is a fairly good approximation to the "average width" of the cube, which is, in fact, $3/2 = 1.5$.)

9. The plane passing through one edge of length a and the perpendicular of length b divides the tetrahedron into two *more accessible* congruent tetrahedra, each with base $ab/2$ and height $a/2$. Hence the required volume

$$= 2 \cdot \frac{1}{3} \cdot \frac{ab}{2} \cdot \frac{a}{2} = \frac{a^2 b}{6}.$$

10. The base of the pyramid is a polygon with n sides. In the case (a) the n lateral edges of the pyramid are equal; in the case (b) the altitudes (drawn from the apex) of its n lateral faces are equal. If we draw the altitude of the pyramid and join its foot to the n vertices of the base in the case (a), but to the feet of the altitudes of the n lateral faces in the case (b), we obtain, in both cases, *n right triangles of which the altitude* (of the pyramid) *is a common side:* I say that these n right triangles are congruent. In fact the hypotenuse [a lateral edge in the case (a), a lateral altitude in the case (b)] is of the same length in each, according to the definitions

laid down in the proposed problem; we have just mentioned that another side (the altitude of the pyramid) and an angle (the right angle) are common to all. In the n congruent triangles the third sides must also be equal; they are drawn from the same point (the foot of the altitude) in the same plane (the base): they form n radii of a circle which is circumscribed about, or inscribed into, the base of the pyramid, in the cases (a) and (b), respectively. [In the case (b) it remains to show, however, that the n radii mentioned are perpendicular to the respective sides of the base; this follows from a well-known theorem of solid geometry on projections.]

It is most remarkable that a plane figure, the isosceles triangle, may have *two different analogues* in solid geometry.

11. Observe that the first equation is so related to the last as the second is to the third: the coefficients on the left-hand sides are the same, but in opposite order, whereas the right-hand sides are opposite. Add the first equation to the last and the second to the third:

$$6(x + u) + 10(y + v) = 0,$$
$$10(x + u) + 10(y + v) = 0.$$

This can be regarded as a system of two linear equations for two unknowns, namely for $x + u$ and $y + v$, and easily yields

$$x + u = 0, \qquad y + v = 0.$$

Substituting $-x$ for u and $-y$ for v in the first two equations of the original system, we find

$$-4x + 4y = 16$$
$$6x - 2y = -16.$$

This is a simple system which yields

$$x = -2, \qquad y = 2, \qquad u = 2, \qquad v = -2$$

12. Between the start and the meeting point each of the friends traveled the same distance. (Remember, distance = velocity $\times$ time.) We distinguish two parts in the condition:

Bob traveled as much as Paul:

$$ct_1 - ct_2 + ct_3 = ct_1 + pt_2 + pt_3,$$

Paul traveled as much as Peter:

$$ct_1 + pt_2 + pt_3 = pt_1 + pt_2 + ct_3.$$

The second equation yields

$$(c - p)t_1 = (c - p)t_3.$$

We assume, of course, that the car travels faster than a pedestrian, $c > p$. It follows

$$t_1 = t_3;$$

that is, Peter walks just as much as Paul. From the first equation, we find that

$$\frac{t_3}{t_2} = \frac{c + p}{c - p}$$

which is, of course, also the value for t_1/t_2. Hence we obtain the answers:

(a) $$\frac{c(t_1 - t_2 + t_3)}{t_1 + t_2 + t_3} = \frac{c(c + 3p)}{3c + p}$$

(b) $$\frac{t_2}{t_1 + t_2 + t_3} = \frac{c - p}{3c + p}$$

(c) In fact, $0 < p < c$. There are two extreme cases:

If $p = 0$ (a) yields $c/3$ and (b) yields $1/3$

If $p = c$ (a) yields c and (b) yields 0.

These results are easy to see without computation.

13. The condition is easily split into four parts expressed by the four equations

$$\begin{aligned} a - d + bg^{-1} &= 85 \\ a + b &= 76 \\ a + d + bg &= 84 \\ 3a &= 126. \end{aligned}$$

The last equation yields $a = 42$, then the second $b = 34$. Adding the remaining two equations (to eliminate d), we obtain

$$2a + b(g^{-1} + g) = 169.$$

Since a and b are already known, we have here a quadratic equation for g. It yields

$$g = 2, \qquad d = -26 \quad \text{or} \quad g = 1/2, \qquad d = 25.$$

The progressions are

$$\begin{matrix} 68, 42, 16 & & 17, 42, 67 \\ & \text{or} & \\ 17, 34, 68 & & 68, 34, 17 \end{matrix}$$

14. If a and $-a$ are the roots having the least absolute value, they will stand next to each other in the progression which will, therefore, be of the form

$$-3a, \ -a, \ a, \ 3a.$$

Hence the left-hand side of the proposed equation must have the form

$$(x^2 - a^2)(x^2 - 9a^2).$$

Carrying out the multiplication and comparing coefficients of like powers, we obtain the system

$$\begin{aligned} 10a^2 &= 3m + 2, \\ 9a^4 &= m^2. \end{aligned}$$

Elimination of a yields

$$19m^2 - 108m - 36 = 0.$$

Hence $m = 6$ or $-6/19$.

15. Let a, b, and c denote the sides, the last being the hypotenuse. The three parts of the condition are expressed by

$$\begin{aligned} a + b + c &= 60 \\ a^2 + b^2 &= c^2 \\ ab &= 12c. \end{aligned}$$

Observing that

$$(a + b)^2 = a^2 + b^2 + 2ab$$

we obtain

$$(60 - c)^2 = c^2 + 24c.$$

Hence $c = 25$ and either $a = 15$, $b = 20$ or $a = 20$, $b = 15$ (no difference for the triangle).

16. The three parts of the condition are expressed by

$$\sin \alpha = \frac{x}{a},$$

$$\sin \beta = \frac{x}{b},$$

$$c^2 = a^2 + b^2 - 2ab \cos \gamma$$

The elimination of a and b yields

$$x^2 = \frac{c^2 \sin^2 \alpha \sin^2 \beta}{\sin^2 \alpha + \sin^2 \beta - 2 \sin \alpha \sin \beta \cos \gamma}.$$

17. We conjecture that

$$\frac{1}{2!} + \frac{2}{3!} + \cdots + \frac{n}{(n+1)!} = 1 - \frac{1}{(n+1)!}.$$

Following the pattern of INDUCTION AND MATHEMATICAL INDUCTION, we ask: Does the conjectured formula remain true when we pass from the value n to the next value $n + 1$? Along with the formula above we should have

$$\frac{1}{2!} + \frac{2}{3!} + \cdots + \frac{n}{(n+1)!} + \frac{n+1}{(n+2)!} = 1 - \frac{1}{(n+2)!}$$

Check this by subtracting from it the former:

$$\frac{n+1}{(n+2)!} = -\frac{1}{(n+2)!} + \frac{1}{(n+1)!}$$

which boils down to

$$\frac{n+2}{(n+2)!} = \frac{1}{(n+1)!}$$

and this last equation is obviously true for $n = 1, 2, 3, \ldots$ hence, by following the pattern referred to above, we can prove our conjecture.

18. In the nth line the right-hand side seems to be n^3 and the left-hand side a sum of n terms. The final term of this sum is the mth odd number, or $2m - 1$, where

$$m = 1 + 2 + 3 + \cdots + n = \frac{n(n+1)}{2};$$

see INDUCTION AND MATHEMATICAL INDUCTION, 4. Hence the final term of the sum on the left-hand side should be

$$2m - 1 = n^2 + n - 1.$$

We can derive hence the initial term of the sum considered in *two* ways: going back $n - 1$ steps from the final term, we find

$$(n^2 + n - 1) - 2(n - 1) = n^2 - n + 1$$

whereas, advancing one step from the final term of the foregoing line, we find

$$[(n-1)^2 + (n-1) - 1] + 2$$

which, after routine simplification, boils down to the same: good! We assert therefore that

$$(n^2 - n + 1) + (n^2 - n + 3) + \cdots + (n^2 + n - 1) = n^3$$

where the left-hand side indicates the sum of n successive terms of an arithmetic progression the difference of which is 2. If the reader knows the rule for the sum of such a progression (arithmetic mean of the initial term and the final term, multiplied by the number of terms), he can verify that

$$\frac{(n^2 - n + 1) + (n^2 + n - 1)}{2} n = n^3$$

and so prove the assertion.

(The rule quoted can be easily proved by a picture little different from Fig. 18.)

19. The length of the perimeter of the regular hexagon with side n is $6n$. Therefore, this perimeter consists of $6n$ boundary lines of length 1 and contains $6n$ vertices. Therefore, in the transition from $n - 1$ to n, V increases by $6n$ units, and so

$$V = 1 + 6(1 + 2 + 3 + \cdots + n) = 3n^2 + 3n + 1;$$

see INDUCTION AND MATHEMATICAL INDUCTION, 4. By 3 diagonals through its center the hexagon is divided into 6 (large) equilateral triangles. By inspection of one of these

$$T = 6(1 + 3 + 5 + \cdots + 2n - 1) = 6n^2$$

(rule for the sum of an arithmetic progression, quoted in the solution of Problem 18). The T triangles have jointly $3T$ sides. In this total $3T$ each internal line of division of length 1 is counted twice, whereas the $6n$ lines along the perimeter of the hexagon are counted but once. Hence

$$2L = 3T + 6n, \qquad L = 9n^2 + 3n.$$

(For the more advanced reader: it follows from Euler's theorem on polyhedra that $T + V = L + 1$. Verify this relation!)

20. Here is a well-ordered array of analogous problems: Compute A_n, B_n, C_n, D_n and E_n. Each of these quantities represents the number of ways to pay the amount of n cents; the difference is in the coins used:

A_n only cents
B_n cents and nickels
C_n cents, nickels, and dimes
D_n cents, nickels, dimes, and quarters
E_n cents, nickels, dimes, quarters, and half dollars.

The symbols E_n (reason now clear) and A_n were used before.

All ways and manners to pay the amount of n cents with the five kinds of coin are enumerated by E_n. We may, however, distinguish two possibilities:

First. No half dollar is used. The number of such ways to pay is D_n, by definition.

Second. A half dollar (possibly more) is used. After the first half dollar is laid on the counter, there remains the amount of $n - 50$ cents to pay, which can be done in exactly E_{n-50} ways.

We infer that

$$E_n = D_n + E_{n-50}.$$

Similarly

$$D_n = C_n + D_{n-25},$$
$$C_n = B_n + C_{n-10},$$
$$B_n = A_n + B_{n-5}.$$

A little attention shows that these formulas remain valid if we set

$$A_0 = B_0 = C_0 = D_0 = E_0 = 1$$

(which obviously makes sense) and regard any one of the quantities A_n, B_n . . . E_n as equal to o when its subscript happens to be negative. (For example, $E_{25} = D_{25}$, as can be seen immediately, and this agrees with our first formula since $E_{25-50} = E_{-25} = 0$.)

Our formulas allow us to compute the quantities considered *recursively*, that is, by going back to lower values of n *or* to former letters of the alphabet. For example, we can compute C_{30} by simple addition if C_{20} and B_{30} are already known. In the table below the initial row, headed by A_n, and the initial column, headed by o, contain only numbers equal to 1. (Why?) Starting from these initial numbers, we compute the others recursively, by simple additions: any other number of the table is equal either to the number above it or to the sum of two numbers: the number above it and another at the proper distance to the left. For example,

$$C_{30} = B_{30} + C_{20} = 7 + 9 = 16$$

The computation is carried through till $E_{50} = 50$: *you can pay* 50 *cents in exactly* 50 *different ways.* Carrying it further, the reader can convince himself that $E_{100} = 292$: *you can change a dollar in* 292 *different ways.*

n	0	5	10	15	20	25	30	35	40	45	50
A_n	1	1	1	1	1	1	1	1	1	1	1
B_n	1	2	3	4	5	6	7	8	9	10	11
C_n	1	2	4	6	9	12	16	20	25	30	36
D_n	1	2	4	6	9	13	18	24	31	39	49
E_n	1	2	4	6	9	13	18	24	31	39	50